Culture
of
Nonsalmonid
Freshwater
Fishes

Editor:

Robert R. Stickney
Professor and Director
School of Fisheries
University of Washington
Seattle, Washington

CRC Press, Inc.
Boca Raton, Florida

Library of Congress Cataloging in Publication Data
Main entry under title:

Culture of nonsalmonid freshwater fishes.

  Bibliography: p.
  Includes index.
  1. Fish-culture.   2. Fishes, Fresh-water.
I. Stickney, Robert R.   II. Title: Nonsalmonid freshwater
fishes.
SH159.C85   1986   639.3′1   85-14954
ISBN 0-8493-6374-8

  Direct all inquiries to CRC Press, Inc., 2000 Corporate Blvd., N.W., Boca Raton, Florida, 33431.

© 1986 by CRC Press, Inc.

International Standard Book Number 0-8493-6374-8

Library of Congress Card Number 85-14954
Printed in the United States

PREFACE

Fish culture has been practiced for many centuries. Both the ancient Chinese and the Pharoahs of Egypt left records which indicate that some type of fish culture has been developed by their civilizations. Yet, until only the past 2 or 3 decades, the culture of nonsalmonid fish could be best classified as an art or an artisinal pursuit, not a science. Researchers involved in fish culture today are often perceived as pseudo-scientists by some elitists among their peers. Yet, the culture on nonsalmonid fish like the channel catfish has developed from an idea in the mid-1950's to a $100 million plus annual business in the 1980's, largely as a result of the efforts of research scientists. Many of these dedicated individuals have focused on fish culture in the traditional agricultural sense; that is, the goal has been to produce food for human consumption. Others, including geneticists, physiologists, endocrinologists, toxicologists, and representatives of many other disciplines, have contributed greatly to fish culture either by conducting directed research or as a result of studies of a more basic scientific nature. The literature cited in this text underwrites the point that contributions to the culture of nonsalmonid fishes have been made by individuals who represent all points on the spectrum from purely basic to highly applied research. The contributions from all of these individuals have been important to our present understanding and to the rapid strides which have been made in the production of nonsalmonid fishes. Admittedly, successful culture of many nonsalmonid species (carp being a notable exception) has lagged a few decades behind that of trout and salmon. In fact, salmonid culturists have shown those of us involved in nonsalmonid culture the paths along which we should travel to solve many of our problems. Yet, salmonid and nonsalmonid fishes are sufficiently different that the knowledge gained from research on either is frequently not directly applicable to the other. Thus, there is some logic in treating nonsalmonid fishes separate from their coldwater relatives.

Nonsalmonid fish culture has moved from the research laboratory into the commercial sector with respect to many species. Others will undoubtedly follow after a few nagging problems are solved. The future for persons interested in both research and production promises to be exciting. There are still species which have received little or no attention but hold potential as culture candidates. Recent developments in molecular biology and genetic engineering, could have immense impacts on future aquacultural development. As large quantities of suitable quality water become increasingly scarce, as is occuring throughout much of the world, new methodology will have to be developed so fish culture will not be eliminated from some areas where it is currently widely practiced. Integration of fish culture into more traditional agricultural pursuits, particularly in developing nations, offers some intriguing potentials as well as problems.

A volume such as this represents only one point along a broad continuum. Between the time that the chapters which follow were written and the text appeared in print, a great deal of additional data were gathered and various publications produced which could obviously not be included. We feel, however, that the information presented is relatively complete to mid-1984.

Robert R. Stickney
Redmond, Washington
September 4, 1985

THE EDITOR

Robert R. Stickney, Ph.D., is Professor and Director of the School of Fisheries, University of Washington, Seattle, Washington.

Dr. Stickney received his B.S. degree in Zoology at the University of Nebraska in 1967, his M.A. in Zoology at the University of Missouri in 1968, and his Ph.D. in Biological Oceanography at Florida State University in 1971. Dr. Stickney served as a Research Associate and Assistant Professor on the staff of the Skidaway Institute of Oceanography from 1971 to 1975, and as Assistant Professor, Associate Professor, and Professor in the Department of Wildlife and Fisheries Sciences at Texas A&M University from 1975 through 1983. He directed the Aquaculture Research Center at Texas A&M University throughout his tenure at that institution. Dr. Stickney became Director of the Fisheries Research Laboratory of Southern Illinois University in January, 1984 and served in that capacity until September, 1985 when he accepted his present position at the University of Washington.

Dr. Stickney's research has centered around aquaculture, with emphasis on warmwater fishes such as channel catfish and tilapia. He has also worked on the culture of flounders, freshwater shrimp, and various other species, and has conducted ecological research along the southeastern coast of the U.S. His major areas of research activity have been in fish nutrition and determination of the environmental requirements of fishes. Research projects have taken him to the Far East, Middle East, and Latin America where he has conducted work in both marine and freshwater environments. Dr. Stickney has written books on aquaculture and estuarine ecology, has published in excess of 70 scientific papers, and has written, co-authored, or edited numerous book chapters, abstracts, and technical reports.

DEDICATION

An old Chinese proverb states that if you give a man a fish he will have food for a day, but if you teach him how to raise fish, he will have food for a lifetime. This book is dedicated to people who produce fish and those future generations who will learn the techniques of fish production.

CONTRIBUTORS

Gary J. Carmichael
Fisheries Biologist
San Marcos National Fish Hatchery
  and Technology Center
San Marcos, Texas

James T. Davis
Fisheries Specialist
Texas Agricultural Extension Service
Texas A&M University
College Station, Texas

Roy C. Heidinger
Acting Director
Fisheries Research Laboratory
Southern Illinois University
Carbondale, Illinois

Terrence B. Kayes
Assistant Director
University of Wisconsin Aquaculture
  Program
Department of Food Science
University of Wisconsin-Madison
Madison, Wisconsin

Jerome Howard Kerby
Assistant Leader
North Carolina Cooperative Fishery
  Research Unit
Associate Professor
Department of Zoology
North Carolina State University
Raleigh, North Carolina

Robert B. McGeachin
Research Scientist
Caribbean Marine Research Center
Riviera Beach, Florida

John G. Nickum
Leader
Iowa Cooperative Fishery Research
  Unit
U.S. Fish and Wildlife Service
Associate Professor
Department of Animal Ecology
Iowa State University
Ames, Iowa
Presently:
Fisheries Biologist
U.S. Fish and Wildlife Service
Washington, D.C.

Bill A. Simco
Professor
Department of Biology
Memphis State University
Memphis, Tennessee

Robert R. Stickney
Professor and Director
School of Fisheries
University of Washington
Seattle, Washington

Joseph R. Tomasso, Jr.
Assistant Professor
Department of Biology
Southwest Texas State University
San Marcos, Texas

Harry Westers
Hatchery Operations Manager
Michigan Department of Natural
  Resources
Fisheries Division
Lansing, Michigan

J. Holt Williamson
Fisheries Biologist
San Marcos National Fish Hatchery
  and Technology Center
U.S. Fish and Wildlife Services
San Marcos, Texas

TABLE OF CONTENTS

Chapter 1

INTRODUCTION

Robert R. Stickney

TABLE OF CONTENTS

I. SCOPE OF THE TEXT

Aquaculture, the rearing of aquatic organisms under controlled, or semicontrolled conditions,[1] has expanded rapidly in the U.S. over the past 2 decades. The discipline includes the culture of both freshwater and marine organisms, with saltwater species under culture falling within the subdiscipline mariculture.

Fish species of freshwater aquaculture interest, the focus of this book, may be produced for direct consumption by man, as species to be stocked for ultimate capture by sport fishermen, or as bait. Another large area of freshwater aquacultural production involves aquarium fishes, popular with home aquarium hobbiests and produced in the U.S. and abroad. Aquarium species are not considered in this volume.

The species under consideration here are produced primarily by commercial aquaculturists or by state and federal hatcheries for stocking in public waters. Each of the states has an agency (Department of Fish and Game, Conservation Commission, Department of Parks and Wildlife, etc.) which stocks fish in public and, in most cases, private waters. In addition, the federal government of the U.S. has several hatcheries which provide fish for stocking into public waters. State, federal, and university laboratories frequently produce fish for research purposes. For most of the discussion which follows, the species under culture are produced by both the private sector and public agencies.

In this text, we consider the culture of selected freshwater fishes outside of the family Salmonidae. Included are both warmwater and or coolwater species. Warmwater species are those which have temperature optima in the vicinity of 26 to 30°C, while coolwater species are best cultured in the 20 to 25°C range. Among the former are such species as channel catfish *(Ictalurus punctatus),* tilapia (*Tilapia* spp.), and carp (various genera and species). The latter group includes yellow perch (*Perca flavescens),* northern pike (*Esox lucius),* muskellunge (*Esox masquinongy),* and walleye (*Stizostedion vitreum).* In addition to the species mentioned, striped bass *(Morone saxitilis)* is included in this text since that species, while anadromous, is widely cultured for stocking into landlocked freshwater environments. We also consider the culture of minnows and goldfish, each of which is utilized as forage for certain carnivorous games species and as bait by sport fishermen. Goldfish, of course, also enjoy popularity among hobbiests.

Various additional species of freshwater fishes are under culture around the world. Included are catfishes of the genus *Clarias* (walking catfish) and such cichlids as the Oscar (*Cichla ocellaris).* Another genus with sportfish potential is *Lates* (Nile perch), with species which reach in excess of 50 kg in some cases. None of these groups of fishes are currently of importance in U.S. aquaculture and thus are not considered in this volume.

The culture techniques involved in the production of carp (particularly common carp, *Cyprinus carpio,* and grass carp, *Ctenopharyngodon idella)* and buffalo (*Ictiobus* spp.), while not currently important in the U.S., are considered because those species are of some interest in this country. Much of the current interest in grass carp and to some extent common carp stems from the controversy which surrounds both species. Buffalo, once considered as a species with good foodfish potential, may reach that status again in the future.

The taxonomy used in this text follows that of the American Fisheries Society.[2] Thus, the various species of tilapia discussed have been given the generic designation *Tilapia,* though Trewavas[3] recently suggested that the genus should be organized into several genera. If her taxonomy is followed, most of the commercially important species of tilapia would be classified as members of the genera *Oreochromis* or *Sarotherodon,* though the common name of most continues to be "tilapia". The suggested taxonomic changes are based primarily upon behavioral differences among the species (e.g., distinctions among genera may be on the basis of whether the species is a mouth-brooder or substrate spawner). Many fisheries scientists do not feel that such behavioral differences are sufficient reason to assign generic status to the species involved, so a minor controversy has developed. Most U.S. aquaculturists continue to refer to the various species of cultured cichlids as *Tilapia,* while many counterparts in Europe and the Middle East have adopted the additional genera *Sarotherodon* and *Oreochromis.*

II. PHILOSOPHY OF COMMERCIAL AQUACULTURE

The goal of commercial foodfish culture in the U. S. and throughout the world is one of achieving a profit. Commercial fish farming, like other forms of agriculture, is aimed at providing food for human populations, but can be successful only if the culturist makes sufficient income to maintain the business and provide a living wage for all persons involved in the enterprise. In most instances, the return from each fish reared is small so a commercial operator must produce large numbers of fish annually to realize the requisite profit. Thus, initial investment and continuing variable costs of operating a fish farm are high. Some of the factors to be considered before becoming involved in a commercial fish farming venture are outlined in Table 1.

While it has been widely claimed, particularly in the popular press, that aquaculture can provide inexpensive animal protein to feed the world, in reality the initial investment and overhead involved in operating a successful fish farm require that the product, no matter what the species, be sold at a price which is normally beyond the means of the truly hungry. Subsistence fish farming, whereby a family or small group of families produces fish for their own consumption and perhaps a few for sale, is widely employed in the developing world. However, competition for land and water along with the low level of intensity which must be employed in such culture operations limits the potential of subsistence aquaculture for providing the quantities of animal protein which the world requires.

A recent development in such developed countries as the U.S. has been growing interest in backyard fish farming. While often advertised as a form of subsistence aquaculture, the types of closed water systems which have been developed for use by the backyard fish farming enthusiast generally require a substantial initial investment and the continuous input of energy. There is no doubt that significant numbers of fish can be grown in backyard systems. Such systems may employ solar heating and may be augmented with facilities for the culture of vegetables using fish waste as a nutrient source. Yet, the suitability of such novelty systems for the general public remains to be demonstrated. Backyard culture systems employ little if any technology which has not been utilized by researchers or commercial aquaculturists for several years unless that technology has been subsequently abandoned as not being cost effective. The approach of this text is to consider more traditional forms of commercial culture and to leave subsistence and backyard fish farming for others to address.

III. CULTURE SYSTEMS

The vast majority of freshwater fishes reared under aquaculture conditions are raised in ponds (Figure 1), though various other types of culture systems are employed in specialized instances. The types of culture systems employed by freshwater aquaculturists can be thought of as lying along a continuum, with lightly stocked farm ponds at one end and closed recirculating systems at the other. Between the two extremes are such cultural strategies as those which employ heavily stocked ponds, cages, linear raceways, circular tanks, or closed systems (Figures 2 through 5).

Table 1
SOME CONSIDERATIONS FOR THE PROSPECTIVE FISH FARMER

Financing   A prospectus should be developed which details the costs of land and capital expenditures for fish stock, buildings, pond construction, operating capital, labor, financing, production, harvesting, marketing, and insurance; the prospectus should contain depreciation schedules and a profit and loss estimate
The prospective fish farmer should be capitalized to the extent of $4500/ha of water to be placed into production
An overhead of up to $250/month for each hectare of water (excluding fingerling costs) can be anticipated

Site   Natural elevations of the selected site should allow for complete drainage of each pond
Water of the proper quality and quantity should be available at reasonable cost
Soils should be analyzed to determine that their physical properties are suitable for water retention
Cores should be taken to determine that desired soil type is available throughout the selected site
Adjacent lands should not be subject to aerial spraying for insects and weeds and your soil should not contain toxic residues
Site should not be subject to flooding
State and federal permits should be obtained in advance of construction
Proper size and shape of ponds should be determined to fit the needs of the planned operation
Potential losses from poaching should be assessed

Fish source   The decision must be made as to whether fish are to be produced on site or purchased from other fish breeders
If breeding is to be conducted on site, sufficient space for maintenance of broodfish and for production of fingerlings must be provided
If fingerlings are to be purchased, the availability of reliable sources of good quality (disease-free) fingerlings must be determined

Feeding   Feed of the proper quality must be available in the required quantity upon demand throughout the growing season
Feeding schedules (usually based upon water temperature) should be formulated and followed
Decisions must be made as to how to adjust feeding rates during the growing season
The culturist must decide whether sinking or floating feed should be used

Harvesting   The most economical harvesting method should be utilized
Special equipment needs for handling fish from harvesting to processing should be assessed
Holding facilities for harvested fish may be required
The need for and size of transportation tanks and trucks should be assessed

Marketing   The proximity to a suitable market should be a factor in site selection
Alternative processing and marketing strategies for the species under culture should be evaluated
Quality control should be exercised to avoid loss of customers as a result of off-flavors or improperly sized fish for the existing market

Management   A trained biologist competent to make diagnosis and initiate treatment for disease should be readily available or on the staff of the fish farm
Suitable laboratory equipment for water quality testing, disease diagnosis, and other routine uses should be available
Emergency backup power should be available in case of electrical failure
The facilities should be designed to provide as much efficiency as possible in day-to-day activities
All-weather access to all facilities should be provided
Areas for expansion should be considered in the initial development phase

Adapted from a checklist prepared by James T. Davis.

As the level of intensity increases (i.e., increasing intensity implies increased density of the culture species per unit area or volume of culture chamber and increasing technological input), the level of management required for successful culture increases. Whereas it may take little or no management to produce a few hundred kilograms per

FIGURE 1. Typical aquaculture ponds.

hectare of fish in a farm pond annually, the production of several thousand kilograms per hectare of fish within a raceway or cage may require the fulltime attention of one or more culturists.

While some differences occur among the various freshwater species under culture, the basic types of culture environments discussed in the following subsections remain relatively constant. Peculiarities associated with the culture of individual organisms are addressed when those species are discussed in later sections and a more complete description of each type of culture unit was provided by Stickney.[1]

A. Ponds

Fish culture ponds range in size from a fraction of a hectare to several hectares. In general, small ponds are utilized by commercial culturists for spawning and fingerling production. Small ponds are also most commonly employed by researchers because of the relative ease and economy with which replicated experiments can be conducted. Production ponds larger than about 10 ha become difficult to manage and do not appear to be in favor among most commercial producers.

Unlike the typical farm pond (Figure 6), culture ponds are usually located in areas which have little relief, are rectangular or square in shape, have well controlled levee and bottom slopes, and do not receive runoff from the surrounding watershed. Culture ponds are most commonly filled with well water, though water can be pumped from an adjacent reservoir or stream. Spring water, including that from geothermal sources, has also been employed. It is important that sufficient water is available to fill all ponds within a reasonable period of time and to maintain the ponds at the desired level plus provide supplemental water in emergencies. While the optimum amount of water availability varies depending upon the source of information, many culturists insist upon having between 100 and 250 ℓ/min of water available for each hectare under culture.

In addition to being able to fill a culture pond rapidly, provision should be made to drain it completely and within a period of no more than about 3 days. Drains may be

FIGURE 2.   Cages.

constructed utilizing gate valves, tilt-over standpipes, or kettles (often referred to as monks) which employ boards to control water level (Figure 7). Drain structures may be very simple, or they may employ a great deal of concrete work in the formation of kettles, catchment basins, and stairways for access and egress. Pond bottoms should slope toward the drain with a grade of about 1%. Side slopes should be 2:1 or 3:1 (for each meter of height, there should be 2 or 3 m of lateral distance). Slopes with those specifications will allow for relatively easy access, will not promote the growth of submerged and emmergent vegetation, and will help reduce erosion problems. Pond levees should be wide enough to mow, and each pond should have at least one side with the levee wide enough to support vehicles used during stocking, feeding, and harvesting. Road levees should be gravel-topped and all levees should be seeded to grass.

Some of the characteristics which a good culture pond should exhibit are summarized in Table 2. Included is the desirability of placing core trenches under perimeter levees, and, if economics permits, beneath all pond levees. A core trench (Figure 8), is a ditch dug to a depth of a meter or so below the elevation of the pond bottom. That ditch is subsequently filled and packed with the same soil used to construct the levee.

FIGURE 3.   Linear concrete raceways utilized for rearing channel catfish.

FIGURE 4.   Circular fiberglass tanks.

FIGURE 5.   A closed water system.

FIGURE 6.   Typical farm pond in south Texas.

The effect is to provide a lock and key arrangement of soil with the same compaction from the top of the levee to well below its base. This minimizes lateral seepage.

The production capabilities of ponds will vary to some extent depending upon the species under culture. At present, channel catfish farmers in Mississippi are producing approximately 4000 kg/ha on the average, up considerably from the production of 1000 to 3000 kg/ha which occurred in most catfish farming areas during the 1960s and

FIGURE 7. Kettle or monk which utilizes two sets of baffle boards. The space between the boards is filled with soil in this case to prevent leakage.

1970s. Even higher levels of production can be achieved if some exchange of water is utilized during the latter part of the growing season, though the cost involved in pumping water may exceed the return obtained from higher standing crops of fish.

B. Closed Systems

Closed water systems feature recirculation of the water with treatment during each passage of that water through the system. The typical closed system features culture chambers, a primary settling chamber for elimination of particulate solids, a biofilter

Table 2
CHARACTERISTICS OF A FISH CULTURE POND

Location	Select land with a gentle slope and lay out ponds to take advantage of existing terrain
Construction	Ponds may be dug into the ground, they may be partly above and partly within the ground or they may be below original ground elevation; slopes and bottom should be well packed during construction to retard erosion and seepage; soil should contain a minimum of 25% clay; rocks, grass, branches, and other foreign objects should be eliminated from levees
Core trenches	A core trench should be dug beneath the perimeter levee of the facility, and it is desirable to provide a core trench under each levee to prevent lateral water seepage
Pond depth	Depth should be 0.5—1.0 m at shallow end, sloping to 1.5—2.0 m at the drain end; deeper ponds may be required in northern regions where the threat of winterkill below deep ice cover exists
Pond configuration	Ponds should be rectangular or square
Side slopes	Construct ponds with 2:1 or 3:1 slopes on all sides
Drain	Gate valves, baffle boards, or tilt-over standpipes should be provided; draining should require no more than 3 days
Inflow lines	Inflow lines should be of sufficient capacity to fill each pond within 3 days; if surface water is used, the incoming water should be filtered to remove undesirable organisms
Total water volume	Sufficient water should be available to fill all ponds on the facility within a few weeks and to maintain them full throughout the growing season; sufficient water should be available to provide 100—250 ℓ/min/ha of total pond area
Levees	Levees should be sufficiently wide to mow; at least one side should be wide enough (2—3 m) to provide vehicular access; road levees should be gravel; grass should be planted on all levees
Orientation	Orient ponds to take advantage of wind mixing, or in areas where wind causes extensive wave erosion, place long axis of pond at right angles to the prevailing wind; use hedge or tree wind breaks as necessary

FIGURE 8.   A core trench has been dug prior to construction of the perimeter levee.

for purification of the water, and a secondary settling chamber or final clarifier to remove particulate matter which escapes from the biofilter (Figure 5).

The biofilter is the heart of any recirculating water system. It provides a large amount of surface area for colonization of the bacteria which detoxify the metabolic wastes produced by the fish in the culture chambers. The primary function of the biofilter is the conversion of ammonia to nitrite and, ultimately nitrate by bacteria of the genera *Nitrosomonas* and *Nitrobacter,* respectively. The reactions are as follows:

$$NH_3 \xrightarrow{\text{\textit{Nitrosomonas}}} NO_2^- \tag{1}$$

$$NO_2^- \xrightarrow{\text{\textit{Nitrobacter}}} NO_3^- \tag{2}$$

Both ammonia and nitrite are extremely toxic to fish.

Sizing of biofilters to provide sufficient handling capacity of fish tanks continues to be more of an art than a science, with total surface area of biofilter medium to culture tank volume varying considerably depending upon the species under culture, biomass within the system, water flow rate, and various other factors. Biofilters should be designed so as not to restrict water flow (avoid sand and gravel filter media, in general). Any material which will support colonization by bacteria may be used (e.g., fiberglass and various types of plastic make good biofilter media. The design and operation of water systems of this type have been discussed by Stickney.[1] Additional information on the engineering aspects of fish culture systems can be found in Wheaton[4] and several of the papers published in the Bio-Engineering Symposium edited by Allen and Kinney.[5]

Closed systems can be utilized to produce extremely high densities of fish. Since the water is being continuously replenished, the fish can be maintained at densities which would equate to something like a million kilograms per hectare in a static pond situation. Because of high energy demands for moving water and often for maintaining water temperature in such systems, they have rarely been proven economical in commercial fish culture, though they have been employed by researchers.

C. Cages

Fish cages (Figure 3) have found a place in culture in instances where fish are reared in streams, reservoirs, lakes, or the intake or discharge canals of power plants. Cages are constructed of rigid frames over which welded wire fabric or nylon netting is stretched and secured. Individual cages may be provided with their own flotation materials, or they may be tied to a floating platform. The major use of cage culture is in bodies of water where the control of fish would be difficult or virtually impossible without some type of close confinement. As long as there is free circulation through each cage and the density of fish does not become too great, water quality within cages is generally good.

Disadvantages of cage culture include difficulty in treating for diseases, the need to provide feed on a daily basis to each individual cage (usually by boat), initial relatively high costs involved in cage construction, need for continuous maintenance, need for replacement at intervals of 5 years or less, and the ease with which fish can be poached. At present, only a small percentage of U.S. commercial fish culturists employ the cage culture technique.

Concentrations of several hundred fish per cubic meter have routinely been maintained in cages by fish culturists. The actual density utilized will depend upon the species under culture and the size to which the fish are to be reared. Good production might be 100 kg/m³ or more.

D. Raceways

Raceways are culture chambers in which water is introduced at a relatively rapid rate so that the water within each unit is replaced at intervals of from a few minutes to a few hours. Because of the rapid turnover rate, the fish are constantly exposed to new water of high quality; thus, it is possible to maintain extremely high densities in small volumes. Raceways may be linear (Figure 4) or circular (Figure 5). They may be constructed so that the effluent water from one is aerated and flowed through one or more subsequent raceways before being discharged from the system, or each raceway may receive its own supply of new water. When used in series, water quality tends to deteriorate as it flows from raceway to raceway, so stocking density may have to be reduced in the downstream units.

Circular raceways are most commonly utilized by researchers, though some commercial operations have also employed them successfully. Circular raceways normally have center drains and are rarely over about 10 m in diameter, whereas linear raceways of as much as 100 m in length have been used, particularly in the trout growing regions of the U.S. Linear raceways are most often constructed of concrete, while circular raceways may be of concrete, fiberglass, or metal. In most cases the water in a raceway is maintained at a depth of about 1 m, though deep circular raceways (several meters in height), called silos, have been developed and can be used for species which will distribute themselves throughout the water column.

The carrying capacity of a raceway will depend upon the flow rate utilized, the species under culture and the quality of the incoming water. In most instances it should be possible to carry as many or more fish per unit volume of water in an open raceway as in a closed system.

IV. WATER QUALITY

A common requirement of each of the species considered below is the need for water of adequate quality. As indicated above, the optimum growth of the various species falls, in general within the range of 20 to 30°C, though most of them can tolerate considerably wider ranges of temperature. Some, such as the striped bass and most species of tilapia, can also tolerate wide ranges of salinity. In freshwater environments, salinity measurement is not necessary, but careful monitoring of temperature may be required as a means of determining when various management strategies should be employed.

Below are brief descriptions of some of the more important water quality variables which should be measured by freshwater fish culturists. Table 3 presents information on the preservation of water samples prior to analysis. Complete details of the chemistry involved with each determination can be found in Boyd.[6]

A. Temperature

Culture system temperature should be routinely monitored since management decisions with respect to such things as spawning and on feeding rates are made based primarily upon that variable. Daily records from at least representative culture chambers should be maintained and, if some culture chambers are known to vary significantly from the mean, those should be monitored individually. If thermal adjustment from ambient water temperature is being undertaken, constant measurement may be desirable.

Temperature can be measured with a glass thermometer, an in-line temperature gauge, or with a thermistor (often used in conjunction with an oxygen electrode) which is utilized in conjuction with an electrically operated metering device. Monitoring may be continuous or at intervals. Recorders are available for producing records of temperature and it is possible to utilize electronic circuitry to perform functions such as shutting off incoming water if temperature fluctuates outside a given range or turning on and off heaters to help maintain the desired temperature.

Table 3
METHODS OF PRESERVATION OF WATER SAMPLES[6]

Determination	Preservative	Holding time
Alkalinity	Cool to 4°C	24 hr
Biochemical oxygen demand	Cool to 4°C	6 hr
Carbon dioxide	Cool to 4°C	2 hr
Chemical oxygen demand	1 mℓ/ℓ H_2SO_4	7 days
Dissolved oxygen	Fix immediately for Winkler titration	6 hr
Hardness	Cool to 4°C or add 1 mℓ HNO_3	7 days
Ammonia	Cool to 4°C or add 1 mℓ H_2SO_4	24 hr
	add 40 mg/ℓ $HgCL_2$	7 days
		7 days
Nitrate	Same as ammonia	
Total phosphorus	Cool to 4°C	7 days
Soluble orthophosphate	Cool to 4°C	2 hr
Settleable solids	—	24 hr
Chlorophyll a	Cool to 4°C	12 hr

In pond systems and other large-scale culture operations it is not generally possible to modify water temperature from ambient unless water supplies of two or more temperatures can be mixed to produce the desired thermal regime. Recirculating systems are usually temperature controlled, often by placing them within a heated building and, in some instances, by providing supplemental water heating. Temperature control in cages, which are normally placed in a large aquatic environment, is often not possible. However, in instances where cage culture is conducted in conjuction with power plant cooling water discharge, it may be possible to move the cages into water of the desired temperature. This technique has received a considerable amount of attention, but problems with gas bubble disease (caused by supersaturation of gases, particularly in winter, upon passage of cold water through power plant condensors) have been severe in many instances where the technique has been employed.

B. Alkalinity

One of the water quality variables which normally requires only infrequent measurement is alkalinity, defined as the total amount of acid required to titrate the bases in a water sample. Included in the total alkalinity of water sample are carbonate, bicarbonate, hydroxide, silicate, phosphate, ammonia, and various organic compounds, but the majority of the bases are attributable to carbonate and bicarbonate, so an alternative definition is that alkalinity is the total bicarbonate and carbonate in a water sample. It is also a measurement of the buffering capacity of the water.

Measurement of alkalinity is by two-step titration. The first step titrates the sample to its phenophthalein endpoint and measures carbonate alkalinity (reported as mg/ℓ $CaCO_3$). The amount of titrant used in the second step, indicated with methyl orange, when added to the first provides a measurement of total alkalinity (also reported as mg/ℓ $CaCO_3$). The difference between total and carbonate alkalinity is a measure of bicarbonate alkalinity.

Most water sources maintain constant or nearly constant alkalinity over long periods of time. Some pond systems have sources of water which are of extremely low alkalinity. When alkalinity is less than about 20 mg/ℓ, it may be necessary to add a source of carbonate to the ponds. This problem and its solution have been discussed by other authors.[6-8]

Closed water systems, which may be initially filled with water of suitable alkalinity often become increasingly acid with time. As organic acids are formed in the system due to the activity of the fish as well as the microflora in the biofilter, carbonate and bicarbonate ions may be depleted and the pH of the water may eventually be reduced, perhaps to critically low levels. The problem can be avoided if excess calcium carbonate is provided in such systems by placing oyster shell or crushed limestone in a settling basin or in the biofilter chamber. The release of carbonate ions ($CO_3^=$), and subsequent formation of bicarbonate ions (HCO_3^-) from the calcium carbonate present in oyster shell or crushed limestone prevents the water from becoming acid. The reactions involved are as follows:

$$CaCO_3 \rightarrow Ca^{++} + CO_3^= \qquad (3)$$

$$CO_3^= + H^+ \rightarrow HCO_3^- \qquad (4)$$

Each of the reactions is reversible and will go in the opposite direction if the pH of the system becomes alkaline.

C. Hardness

The concentration of calcium plus magnesium expressed as equivalent $CaCO_3$ is a measure of total water hardness. Water hardness does not normally vary temporally, but in some instances where the water is extremely soft (less than about 20 mg/ℓ total hardness), it may be necessary to add gypsum or some other source of divalent cations to the water.[6]

D. Ammonia

Measurement techniques for ammonia such as the colorimetric test using Nessler's reagent and the ammonia-specific electrode method lead to the determination of total ammonia. Ionized ammonia accumulates in the water as a result of the elimination of amonium ion (NH_4^+) through the gills[9] by fishes within the culture system. As indicated in Equations 1 and 2, ammonia can be converted to less toxic nitrate through a nitrite intermediate.

Total ammonia concentration is made up of both ionized and unionized ammonia. The latter, NH_3, is the more toxic form.[10-12] The percentage of total ammonia contributed by the un-ionized form is affected primarily by temperature and pH (although other aspects of water quality also exert some influence on the relationship). As temperature and pH increase, so does the percentage of unionized ammonia in the sample. In order to determine the percentage of unionized ammonia in a given sample, it is necessary to make measurements of other aspects of water quality and calculate the relationship between the two forms of ammonia. Tables of percentage of unionized ammonia values have been published,[6,13] but, in general, if total ammonia remains below 1.0 mg/ℓ in water of normal temperature and at a pH which is no more than slightly alkaline, problems with ammonia toxicity are generally avoided. Exposure to ammonia may lead to the development of resistance to the chemical,[14] at least in *Tilapia aurea*.

E. Nitrite

Nitrite toxicity has not historically been an important problem in ponds and other types of water systems other than those of the recirculating type.[1] Fishes suffering from nitrite toxicity have been frequently observed in recirculating systems in which the bacterial flora has not been adequately established on the biofilter prior to fish stocking. Fishes suffering from nitrite toxicity exhibit chocolate-colored blood, hence the common name, brown blood disease. When nitrite levels become sufficiently high, hemoglobin will react with nitrite to form a compound known as methemoglobin. The latter compound is unable to carry oxygen via the blood to the various tissues of the body, so the fish may eventually become asphyxiated. Fish suffering from methemoglobinemia may begin to swim erratically just prior to dying.[15]

With the practice of growing channel catfish to densities in excess of 4000 kg/ha during the past few years, methemoglobinemia has begun to appear in ponds, particularly during the latter part of the growing season. Environmental chloride has been shown to inhibit methemoglobin formation in fish and can, therefore, be utilized to protect them from nitrite toxicity.[16] Such chemicals as calcium chloride and sodium chloride have been added to ponds exhibiting nitrite toxicity by commercial producers. Levels of either chemical of 60 mg/ℓ have been shown to significantly enhance the survival of channel catfish exposed to toxic levels of nitrite.[16]

Median tolerance limit (TL_m) values (the concentration of a substance which leads to 50% mortality of the exposed organism) for channel catfish exposed to nitrite have been elaborated by Konikoff.[15] The TL_m values after 24, 48, 72, and 96 hr were 33.8, 28.8, 27.3, and 24.8 mg/ℓ, respectively. All tests were conducted with 40 g fish at 21°C.

F. Dissolved Oxygen

Perhaps the most critical water quality variable over which the culturist has at least some measure of control is dissolved oxygen (DO). Unlike temperature, DO is largely under biological control. Oxygen can be dissolved in water through diffusion from the atmosphere (sometimes aided by mechanical forces) and through photosynthetic activity. Diurnal fluctuations in DO are largely controlled by the latter mechanism.

In pond systems, DO levels are typically lowest near dawn as a result of continuous respiratory demand during the night. As photosynthetic oxygen production begins to place more oxygen in the water than is being removed by respiration, the DO level begins to increase. This increase usually continues throughout the daylight hours, with another decrease in DO beginning about dusk. Net 24-hr gains or losses in DO level are influenced by any activity which affects photosynthesis, respiration, or mechanical mixing. Some of the factors which can lead to a net reduction in DO are prolonged periods of overcast weather, increased turbidity, and the presence of herbicides in the water. All have a negative effect on phytoplankton productivity.

For virtually all warmwater fish species, it is desirable to maintain a DO level of 5 mg/ℓ, though many species such as channel catfish and tilapia can tolerate lower DOs. Stress is certainly indicated for most species when DO falls below 3.0 mg/ℓ, and death will eventually occur at even lower DO levels. The conscientious fish farmer will routinely monitor DO at least in the early morning during warm weather when fish densities and standing crops are high. Many farmers monitor throughout the night, utilizing personnel to collect data or employing automated systems of data gathering. Prediction of nighttime DO depletions is possible,[17] but not completely reliable as yet, so careful monitoring continues to be important.

Emergency aeration procedures for overcoming low DO[18] include paddlewheel aerators (Figure 9), pumping new water, recirculating pond water, injecting compressed air or air provided by low pressure blowers, and the employment of agitators. Aeration devices may be utilized routinely at night[19] or as needed when an emergency situation is developing or has occurred. In most instances only a small percentage of the ponds on a given farm will require emergency aeration at a given time.

DO can be measured chemically through a method known as the Winkler titration. Oxygen meters, commonly used in conjunction with a thermistor to determine temperature, are also widely used.[6] The latter devices are normally compensated for altitude, salinity, and temperature (as each increases, the oxygen-holding capacity of the water is reduced) and can be connected to emergency aerating equipment through microprocessors which will command the aeration device to become activated when the DO reaches a preset lower limit.

FIGURE 9.   A paddlewheel aerator being utilized in a small research pond.

V. NUTRITION

Cultured fish species may be provided with prepared diets or natural productivity may be stimulated, most commonly through fertilization, to provide increased amounts of natural food organisms. In either case, the goal of the culturist is to provide sufficient food of the proper quality to promote rapid growth in the target fish species. In natural waters the carrying capacity of the system may be quite low (in the vicinity of 140 kg/ha). If both fertilization and supplemental feeding are employed, standing crops of 7500 kg/ha or more can be realized.[20]

Most of the species discussed in this text can be reared at relatively high densities if the water is properly fertilized. The quality and quantity of fertilizer utilized will be dependent upon the type of food organisms being promoted. For example, inorganic fertilizer is often utilized to support blooms of unicellular algae, while zooplankton blooms may be best supported by a fertilization regime which employs organic compounds. Methods and frequency of fertilization for various culture strategies are discussed in the appropriate sections below.

While some fishes do not respond well to prepared feeds, others will accept them from the time of first feeding. In the former group are many of the coolwater fishes, while such fishes as channel catfish and tilapia fall into the latter category. Training to prepared feeds may be required in some instances. In others, fisheries science has not yet solved the problems of attraction, texture, color, and presentation sufficiently to provide a prepared diet which is acceptable to the culture species. Recent breakthroughs have been encouraging and there is no reason to believe that it will not someday be possible to employ prepared feeds for virtually any of the fish species presently under culture.

The specific nutrient requirements of the various fishes discussed in this book may or may not be known with any degree of precision. Some, like the common carp and channel catfish, have been widely researched,[21] while others have received only cursory attention to date. In order for fish nutritionists to define the nutritional requirements of a given species with any degree of confidence, they must evaluate such things as the requirements each species exhibits for proteins and their respective amino acids, energy, lipids, carbohydrates, vitamins, and minerals. In addition, as indicated above, the formulated feed must also be presented in a form and manner acceptable to the species under culture.

The general nutritional requirements of many of the fishes we discuss appear to be similar, but sufficient individual differences exist and broad generalizations cannot be made with any degree of confidence. Thus, we have elected to discuss the nutritional requirements of the various species within the sections which address those organisms.

VI. GENETICS AND SELECTIVE BREEDING

There can be virtually no improvement in culture stocks if the life cycle of the fish species under culture is not under the complete control of the producer. If the culturist is required to go out into nature for broodfish, there is little, if any, opportunity to practice selective breeding. For most of the species discussed below, the production of numerous generations in captivity has become a reality. Yet, for others such as striped bass and some of the coolwater species, it continues to be common practice to collect broodfish from nature. For sportfishes, that approach may be not only the most practical, but also desirable since development of a domestic strain is not the goal of the culturist. On the other hand, selective breeding of sportfishes could lead to a better fighting fish, one which is better able to survive the rigors of nature and, perhaps one which is either more easily caught or more difficult to catch.

With respect to foodfishes, a major goal is often to achieve more rapid growth. Higher dressout percentage and increased fecundity may also be goals of fish breeders. To date, little progress has been made in these areas. One fear is that selection for rapid growth may, in fact, be selection for aggressiveness.

Presently, research and development in the areas of selective breeding and genetics of cultured fishes must be considered as infant disciplines. Some active programs are in existence, but, in general, facilities are limited and the time required for success is long; thus funding is difficult to obtain. With the development of genetic engineering in the past few years, some of the seemingly unattainable goals of fish breeders may soon become a reality. Genetic engineering could provide an entirely new dimension for future fish culturists and may reduce the extent of facilities required to achieve breakthroughs in improved fish production.

VII. RECORD KEEPING

One of the facets of the fish farming business which is often overlooked is the need to keep detailed records of all activities, from amounts of feed utilized to detailed data on the water quality in each pond. Such records help the culturist in the determination of the patterns which frequently become established in a given pond, provide documentation on the mean annual times of spawning and harvest, and can be a valuable aid in overall farm management. In addition, it is necessary to maintain a complete set of records on stocking dates, densities, and sizes. Microcomputer programs have been developed or are under development in many regions of the U.S. to assist the fish farmer in such record keeping chores. Some programs predict yields and feed consumption.[22,23]

VIII. QUALITY OF THE FINAL PRODUCT

Many fish culturists consider their responsibility to be terminated once the fish have been harvested and either sold to a live hauler, a processor, or a distributor. In reality, the fish farmer should be concerned about the product until it not only reaches, but is utilized, by the consumer. In the case of the food fish producer, interest in and responsibility for the product should not end until the fish is served by the housewife or restauranteur. If the fish has an off-flavor, the producer will have lost a customer. Similarly, the bait producer should be concerned that the products produced reach the minnow dealer and, ultimately, the sportsman in good condition. As with any other business, only the satisfied customer will return to the product.

Chapter 2

CHANNEL CATFISH

Robert R. Stickney

TABLE OF CONTENTS

I. INTRODUCTION

The channel catfish (*Ictalurus punctatus*) is currently the most widely cultured warmwater foodfish in the U.S. The species was originally native to the states bordering the Gulf of Mexico and the states within the Mississippi Valley, but it has been widely introduced throughout the U.S. during the past century. The channel catfish is also native to Mexico. In addition to its use in commercial foodfish culture, the species is widely produced in state and federal hatcheries for use in sport fishery stocking programs.

The natural habitat of channel catfish is moderate to swift flowing streams, as well as in lakes and, occasionally, sluggish streams. Because of widespread stocking programs, channel catfish can presently be found in most types of freshwater environments in the U.S.

Other species within the family Ictaluridae which have received the attention of fish culturists are the blue catfish (*I. furcatus*) and the white catfish (*I. catus*). Both species and their hybrids, along with hybrids between them and channel catfish have been produced and in some cases reared commercially, but the bulk of commercial culture continues to be centered on the channel catfish. Little consideration has been given to the various species within the genus known collectively as bullheads, though they could conceivably play a role in fish culture. Another related species, the flathead catfish (*Pylodictis olivaris*), also has not received much attention from fish culturists, largely because it is highly cannibalistic.

II. DEVELOPMENT OF THE COMMERCIAL CATFISH INDUSTRY

Channel catfish have long enjoyed popularity as a foodfish in the southern portion of their range, and the development of commercial production was initiated in the southern U.S. During the late 1950s, Dr. H. S. Swingle of Auburn University in Alabama was the first to demonstrate that channel catfish could be reared in ponds and sold for a profit.[24,25] Since then, the industry has progressed to the point that it now involves revenues to producers of within the range of $100 million annually. Expansion of the industry is evidenced by the fact that in 1963 there were some 960 ha of catfish in production within the U.S., while the figure for 1969 was over 16,000 ha.[26]

Much of the success of the catfish industry can be attributed to active research programs by federal and state governmental agencies as well as various universities. Early work with the species led to the development of suitable prepared feeds, the outline of water quality requirements, and the recognition and development of control measures for the diseases to which catfish are susceptible. The need to reproduce channel catfish for sport stocking led to the development of spawning technology for channel catfish in state and federal hatcheries several decades before commercial foodfish production became a reality. Employment of many of the techniques of fish culture which had been developed for such fishes as trout, salmon, and carp undoubtedly contributed to the rapid development of the commercial catfish industry. In recent years, increased efforts aimed at quality control of the product and an active marketing program have led to significant expansion of the industry. At present, farm-raised channel catfish can be purchased in most states. The product is marketed in various forms and through most retail outlets which handle fishery products, particularly restaurants and supermarkets.

Mississippi has become by far the leading catfish-producing state. By 1982, that state alone had some 25,000 ha under production with a value of over $90 million.[27,28] A survey of Mississippi catfish farmers published in 1980[29] showed that most operations were larger than 45.5 ha (Figure 1). Average stocking density was over 7000 fish/ha (presently over 8000/ha).[30]

FIGURE 1.   Aerial view of part of a commercial catfish farm Delta region of Mississippi (Courtesy of Thomas Wellborn).

Focusing of the industry on the delta region of Mississippi came about because of the abundance of suitable land (flat and of the proper soil type) and the availability of plentiful supplies of suitable water at relatively shallow depth. Initially, much of the industry was centered in Arkansas. Louisiana and Alabama were also leaders in the development of catfish farming, with such states as Texas, California, Georgia, Florida, and a few others being involved to a lesser degree. Isolated catfish producers can also be found in such unlikely states as Illinois and Idaho where the growing season is significantly shorter than that enjoyed by the southern tier of states. However, in at least one case in Idaho, the presence of geothermal water allows for catfish rearing under nearly optimum temperature conditions throughout the year.

III. CULTURE SYSTEMS

While most production of channel catfish is confined to earthen ponds,[26] cages, raceways, and tanks have also been employed by commercial operators. The latter types of systems have been most widely utilized by researchers since replication is convenient and the culture environment can be readily controlled by the researcher.

A. Recirculating and Flow-Through Tank Systems

Prior to the first oil crisis of the early 1970s, it appeared as though a catfish producer could afford to heat water a few degrees and pass it through tanks or raceways of fish and down the drain without recirculation and still make a profit. With the current high cost of energy, such systems no longer hold much attraction, though there continues to be some interest in the development of recirculating systems which conserve not only water, but also heat. The key to a properly functioning recirculating system lies in the design and operating characteristics of the biofilter. Systems have been constructed for catfish which employ updraft, downdraft, submerged, and rotating disc filters (Figures

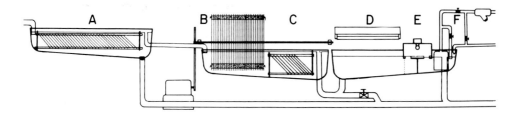

FIGURE 2.   Major components of a recirculating water system. The system is composed of a primary settling chamber containing an inclined plate separator (A), a chamber containing a rotating disc biofilter (B), and a secondary settling chamber with inclined plate separator (C), a plant growing area under a grow-light bulb (D), and a region where the water is exposed to an UV light (E) to kill circulating microorganisms. After leaving the UV light region, the water is pumped through pipes (F) into circular tanks or aquaria (not shown).

FIGURE 3.   Photograph of a portion of the system diagramed in Figure 2. Shown here are the primary settling chamber (lower right), the biofilter chamber (middle), and the third tank which contains the plant growing chamber and the UV light apparatus (upper left).

2 and 3).[1,31-34] An additional step is to integrate hydroponic vegetable production into such systems, thereby reducing the levels of circulating nutrients.[35,36]

Researchers at the Aquaculture Research Center at Texas A&M University have developed a recirculating system which is designed to efficiently maintain water temperature during the winter. Small amounts of discharge water (a few liters per minute) are passed through a chiller which reduces the temperature of the effluent water several degrees Celsius. Freon heated in the chiller is pumped into a heat exchanger where the hot gas transfers heat energy into a storage tank of water. When culture tanks associated with the system fall below a set temperature, an electronic controller signals a

FIGURE 4. Replicated aquaria (in this case each has a capacity of approximately 40 ℓ) are widely employed by catfish researchers. Exchange rates are typically 1 to 2 hr and supplemental aeration is usually provided.

valve to allow water from the storage tank to mix with the recirculating ambient temperature water, thereby increasing the temperature of the system. In essence, the system maintains temperature by removing heat from a source of discharge water and returning that heat to the water within the recirculating system. The water which is cooled by passage through the chiller could be utilized to maintain coldwater or mid-range fish species.

Controlled experimentation with channel catfish was enhanced greatly by the employment of replicated flow-through tanks as experimental units. Maintenance of adequate flows of high quality water in such tank systems makes it possible to hold high densities of fish in small volume experimental units. The pioneering work with large fiberglass circular tank systems was initiated by Andrews and colleagues in Georgia.[37] Most catfish researchers presently employ tank systems of one type or another, ranging from standard glass aquaria to circular polypropylene or fiberglass culture tanks (Figure 4).

B. Cages

The use of cages by catfish farmers (Figure 2, Chapter 1) is limited, but commercial producers have employed the technique in instances where fish are reared under special circumstances. For example, the cage culture technique might be employed in a body of water which cannot be seined easily, in lakes or streams where space is leased from a state (such as is allowed in Arkansas), or in the intake or discharge canals of power plants.[38,39] Cages have also been utilized by researchers for replicated experiments. Stocking density is often several hundred fish per cubic meter, with $500/m^3$ apparently being within the optimum range.[40]

Problems associated with cage culture include difficulty in treating for diseases, loss of fish in the event of cage damage, and the need to provide continuous security to prevent poaching. The major advantage is ease of harvest. The density of cages within a given water body will vary considerably based upon ambient water quality and the rate at which the water within each cage is exchanged.

C. Raceways

Raceways have long been used for trout culture, but are relatively rarely utilized for the production of warmwater fishes. An exception would be a situation in which geothermal water is available at or near the surface and can be obtained with little or no pumping cost. In one such commercial operation (Figure 3, Chapter 1), concrete raceways receive 26.5 m³/min and support densities of 200 to 400 kg/m³ of channel catfish.[41] In that system, dissolved oxygen and ammonia have been found to be the principal limiting factors.

Another possibility is the use of raceways which utilize a large impoundment as a water source and receive water which is subsequently recirculated through the reservoir. One such system was established for research purposes in Tennessee.[42] A series of 33-m long raceways were stocked with 2000 channel catfish in each section. Gradual deterioration of water quality occurred down the progression from the upper to the lower raceway in each series but average production was over 1000 kg per section.

IV. POLYCULTURE

The bulk of the channel catfish industry is based upon monoculture, the rearing of catfish without the presence of other species. However, the concept of polyculture, where two or more compatible species are reared within the same water system, has been utilized with channel catfish as one of the polycultured organisms. Among the animals which have been cultured with channel catfish are buffalo, paddlefish, common carp, grass carp, sunfish, striped bass, golden shiners, tilapia, blue catfish, white catfish, mullet, crayfish, and freshwater shrimp.[43,54] The general conclusion which has been drawn from such stockings is that catfish production was not altered, and total production was augmented by the increases attributable to the other species introduced into the system. In some cases, the presence of other species may actually stimulate feeding by the catfish.[44]

It has been shown that stocking *Tilapia aurea* at densities equal to or greater than 25% of the weight of channel catfish in a flow-through laboratory system is impractical[45] and that in ponds, combinations of catfish and tilapia do not appear to be particularly desirable.[47] Combinations which have included *T. aurea*, carp, and hybrid buffalo have been shown to reduce net yields of channel catfish.[50] However, the stocking of channel catfish in combination with hybrid tilapia (*T. mossambica* male x *T. hornorum* female) in ponds did not affect channel catfish growth,[55] though such water quality variables as turbidity, pH and chlorophyll *a* level were increased by the presence of the tilapia. Thus, the commercial producer should evaluate a given polyculture combination before implementing a large-scale program using the technique.

Polyculture has the major disadvantage that the species produced must be separated for marketing. This is labor intensive and the excessive handling involved can lead to unacceptably high mortalities. In addition, markets must be established for each species produced.

The most commonly cited advantage of polyculture in addition to increased overall production involves the fact that properly selected, the polyculture species can utilize virtually all levels of the natural food chain in a pond. In addition, one or more of the species under polyculture can be offered prepared feed.

Polyculture may also be employed as a management tool. Perhaps the best example is the introduction of grass carp into catfish ponds to control macrophytes and filamentous algae.[53,54] Studies have shown that the grass carp do not prey upon channel catfish or other desirable species.[46]

V. STOCKING DENSITY

Normal strategy for rearing of channel catfish is to stock fry at high densities and rear them to selected fingerling size during the first growing season. Thereafter, the fish are restocked at lower density for growout to market size. Fingerlings from 5 to 10 cm can be produced in about 120 days if fry are stocked at 250,000 to 375,000/ha. If 20-cm fingerlings are desired over the same growing period, fry should be stocked at from 35,000 to 50,000/ha.[26] Fry catfish can also be reared in troughs or tanks[56,57] for later stocking into intensive or extensive growout systems.

Pond growout stocking densities vary widely, with final harvest biomass generally ranging from 1500 to 4000 kg/ha/year (stocking rates may range from 2000 to in excess of 20,000 fish/ha). Most producers desire fish of 0.4 to 0.5 kg, though that varies from one place to another depending upon individual markets. For example, some markets demand fish of nearly 1 kg, while others may request fish smaller than 0.4 kg live weight.

Most producers in Mississippi, and many in other states utilize the continuous harvest technique whereby marketable fish are removed at intervals of a few weeks throughout the year. Periodically, additional fingerlings are stocked to replace harvested fish. Thus, several size classes of catfish may be present in a given pond simultaneously.

Stocking densities in cages have been discussed above and some of the relationships between stocking density and various types of intensive culture systems as well as ponds are more fully developed in the water quality section below. In brief, neither cage size nor fish density appear to affect growth or food conversion efficiency if water quality is maintained.[40,58,59] The same applies to tank rearing.[60] In both cases, water exchange rate controls, at least in part, maximum fish production.

VI. WATER QUALITY

There are virtually thousands of water quality variables which impact upon the performance of fish under culture, but in most circumstances, only a few are monitored since they can be considered to be of greatest importance. The water quality variables discussed below with respect to channel catfish are also among the most important ones for the other species considered in this volume.

A. Temperature

Channel catfish achieve optimum growth at temperatures between 26 and 30°C. Substantial gains can be attained at temperatures from 20 to 34°C.[61-63] Even though a relatively narrow optimum range exists, the species can withstand a wide range of temperatures, as indicated from the fact that it occurs in both subtropical and temperate regions of the North American continent. Feeding activity slows when channel catfish are exposed to temperatures above or below their optimum. The relationship between optimum feeding rate and temperature is elaborated in the section on nutrition below.

B. Dissolved Oxygen

It is generally accepted that dissolved oxygen (DO) concentrations of 5.0 mg/ℓ or above are sufficient to support optimum growth of fishes. Channel catfish are somewhat more tolerant of low DO than some other species, but producers and researchers alike commonly agree that levels below 3.0 mg/ℓ are considered undesirable. At least one study has shown that food consumption and growth are reduced in pond-reared channel catfish exposed to a mean constant 3.5 mg/ℓ DO or less.[64] In tanks, a minimum of 3.0 mg/ℓ DO has been recommended.[65]

Table 1
COMPARISON OF MORTALITY
PERCENTAGES FOR CHANNEL
CATFISH AND CHANNEL CATFISH
X BLUE CATFISH HYBRIDS
EXPOSED TO DO LEVELS OF <1.0
mg/ℓ

	Mortality (%)	
Environment	Channel catfish	Hybrid catfish
Ponds	50.5	7.5
Cages	87.5	51.0
Tanks	100.0	33.0

From Dunham, R. A., *Prog. Fish-Cult.*, 45, 55, 1983.

The saturation DO level of water will vary considerably in relation to temperature, salinity, and altitude. Saturation DO is reduced as any other of three variables increases. Andrews et al.[66] evaluated the effect of various levels of oxygen saturation on growth of channel catfish and found that food consumption and food conversion efficiency were reduced considerably at 36% saturation (those authors compared groups of fish at 36, 60, and 100% saturation).

Just as unrelated fish species sometimes exhibit differences with respect to low oxygen tolerance, such differences can also sometimes be demonstrated in closely related species and even their hybrids (Table 1). The apparently increased ability of blue catfish x channel catfish hybrids to tolerate exposure to low DO[67] was suspected as being due to differences in the ability of hemoglobin to transport oxygen in the blood of the two types of fish. It was recommended that hybrid catfish be used in culture to avoid oxygen depletion-related mortality.

Standard practice in the industry is to obtain DO data from ponds at about sunrise during periods when problems are anticipated. Periods of cloudy, calm weather often lead to oxygen depletions since photosynthesis is limited at such times. The late summer is a period when oxygen depletions are common since the majority of the fish have attained or are approaching harvest size, densities are high, feeding rates are at or near maximum, and the water is quite warm. During the summer, many producers have crews available to obtain DO readings periodically throughout the night. When the level falls below a given minimum, emergency aeration procedures are put into effect. Paddlewheel aerators which run off the power takeoffs of tractors or which are electrically operated (Figure 9, Chapter 1) are commonly utilized by catfish farmers and are effective in overcoming oxygen depletions.[68]

Aeration of catfish ponds has been shown to increase production,[69] particularly when done on a nightly basis in heavily stocked ponds.[70]

Certain management techniques can influence subsequent levels of pond DO. For example, it has been shown that application of the herbicide simazine at 13.4 kg/ha to the bottom of channel catfish ponds before flooding can result in extended periods of low DO and reduction in fish yields and lower food conversion ratios as compared with untreated ponds.[71]

Measurement of DO throughout the night or only near dawn is somewhat labor intensive. Thus, investigators have attempted to relate DO changes to various limnological variables. Variations in DO have been shown to relate to solar radiation, the concentration of chlorophyll *a* and percent oxygen saturation at dawn as well as to

Secchi disc visibility.[72] Based on such findings, Boyd et al.[73] developed a computer program to predict early morning DO levels in catfish ponds. Predictions can be obtained from input data on plankton oxygen consumption, COD, and temperature. In instances where plankton represent the major source of turbidity, Secchi disk visibility may be used to estimate COD or oxygen consumption.[73] The same authors presented a simple graphical technique for estimating dawn DO. That technique involved obtaining a DO value at dusk and another 2 to 3 hr later. Plotting the two points on graph paper and extrapolating to dawn seemed to give a fairly good approximation of what the early morning DO level would be.

C. Ammonia

Chronic growth depression and mortality associated with elevated ammonia concentrations are more commonly reported for intensive culture systems such as tanks and raceways than from pond systems. Yet, with increased stocking rates in ponds in recent years, decreased growth in ponds as a result of high concentrations of ammonia may be on the increase. Most studies of the effects of ammonia concentration on channel catfish growth and survival have been conducted under highly controlled laboratory conditions, though the results of those studies should be applicable to pond production facilities.

The toxicity of unionized and total ammonia on channel catfish has been evaluated only in a few instances.[73-76] Robinette[73] determined that the level of unionized ammonia which would lead to 50% mortality in channel catfish over 24 hr was 2.36 mg/ℓ. Growth was impaired at unionized ammonia concentrations of 0.12 mg/ℓ or higher. The previously accepted concentration of unionized ammonia which causes growth reduction in catfish has recently been challenged by Mitchell and Cech,[77] who determined that residual chloramine levels in charcoal-dechlorinated municipal water will have a synergistic effect with ammonia on the response of catfish. Thus, previous studies may have led to misinterpretation of the effects of ammonia on catfish in instances where monochloramine was present in the test water. In any case, relatively low levels of unionized ammonia in culture water can lead to growth depression.

As the level of ammonia in water increases, growth decreases in a linear fashion.[78] Such other water quality factors as pH, hardness, and temperature also affect ammonia toxicity.[76] Tomasso et al.[79] suggested that elevated pH should be prevented in culture systems to reduce the percentage of unionized ammonia. In most pond and raceway[80] situations, ammonia does not lead to significant problems for fish culturists.

D. Nitrite

The condition known as brown blood disease in channel catfish has been found in association with exposure of the fish to elevated environmental levels of nitrite. When exposed to nitrite, hemoglobin in the blood of the fish is converted to methemoglobin, which gives the blood a chocolate brown color and interferes with the oxygen-carrying capacity of the erythrocytes. When methemoglobinemia becomes sufficiently severe, death from asphyxiation may result. Early reports of methemoglobinemia came only from intensive culture systems wherein flushing rates were inadequate or biofilters were not functioning properly; however, in recent years, the increased stocking densities used by pond culturists, particularly in Mississippi, have led to development of the problem in that type of environment.

Half of the 40-g channel catfish fingerlings exposed to a nitrite concentration of 33.8 mg/ℓ at 21°C will die within 24 hr, with levels of 28.8, 27.3, and 24.8 mg/ℓ nitrite leading to death in 48, 72, and 96 hr, respectively.[81] Exposure to 5 mg/ℓ nitrite for 5 hr will result in a 42.5% level of methemoglobin when the water contains a low level of chloride ion.[82] After 24-hr exposure to the same nitrite level, a methemoglobin level of 90% is reached.[83] Even a nitrite level as low as 1 mg/ℓ will lead to 35% methemoglobin formation after 24 hr.[83] While low levels of nitrite are not considered to be directly lethal, some increased mortality has been shown to occur over a 31-day study period in catfish exposed to 3.7 mg/ℓ nitrite, while growth was retarded at 1.6 mg/ℓ and above.[84]

In order to effectively treat methemoglobinemia, the percentage of methemoglobin in the blood should be known. A sample of about ten fish from a given pond will generally be adequate for making an estimate of the average methemoglobin level in a fish population.[85] Chloride has been shown to be an effective chemical in preventing and treating the problem.[16,79,82,83] The ion may be added as calcium chloride, potassium chloride, or sodium chloride,[83] though the degree of protection varies from one chemical to another. A ratio of 16 chloride to 1 nitrite ion has been shown to completely repress nitrite-induced methemoglobin formation.[82] Sodium bicarbonate also appears to suppress methemoglobin formation,[83] but sodium sulfate has no protective effect.

E. Salinity

The abundance of brackish water in many coastal and even some inland areas has led aquaculturists to experiment with the utilization of that water for the culture of both marine and freshwater fish species. Perry and Avault[86] found that blue, channel, and white catfish could all be successfully reared in salinities of 8 parts per thousand (ppt). The fish could tolerate salinities of as high as 11 ppt, but growth appeared to be reduced. Both blue and channel catfish have been collected from natural waters of slightly higher than 11 ppt salinity.[87]

Some pronounced benefit can be gained by exposing catfish to low salt levels. The use of sodium chloride concentrations of around 2 ppt in catfish culture water has been shown to protect the first against the parasite *Ichthyophthirius multifiliis*[88,89] and may even enhance growth.[90]

Channel catfish eggs can tolerate up to 16 ppt,[91] though tolerance declines to 8 ppt at the time of hatching. Once yolk sac absorption has occurred, salinity tolerance increases to 9 or 10 ppt and will further increase to 11 or 12 ppt by the time the fish are 5 to 6 months of age.[91] Blue catfish appear to be slightly more salinity tolerant than channel catfish,[92] but both species can survive for several days at 14 ppt, as can hybrids among blue, channel, and white catfish.[93] Currently, there is not any emphasis being placed upon the rearing of any catfish species in brackish water within the U.S.

F. pH Control

Pond water which has high alkalinity and low hardness can develop high pH following fertilization.[94] As a means of reducing pH in such situations, hardness may be increased or alkalinity reduced. To increase hardness, agricultural gypsum may be added at the rate of 2000 mg/ℓ without apparent harm to fish.[95] Mandal and Boyd[96] recommended application of agricultural gypsum at four times the difference between total alkalinity and total hardness (in mg/ℓ). Alum (aluminum sulfate) has been used to reduce pond alkalinity.[94] About 1 mg/ℓ of alum will reduce calcium carbonate alkalinity by 1 mg/ℓ.

VII. REPRODUCTION AND GENETICS

Many commercial catfish farmers maintain their own brood stock and produce fingerlings. Genetic research with channel catfish, however, has only received attention in recent years and major advances in genetic improvement of stocks have yet to be made. In the commercial sector, selective breeding generally involves retaining the fish which perform best, which is often interpreted as meaning those which grow the most rapidly.

FIGURE 5. Selection of channel catfish broodstock during the spring is followed by stocking appropriate numbers of each sex in spawning ponds.

A. Reproduction

Reproductive activities with channel catfish include selection of broodstock, establishment of proper conditions for spawning, hatching of the eggs, preparation of ponds, and stocking of the fry. General requirements for spawning of catfish include water temperatures in the range of 21 to 29°C,[26,97,98] with the optimum being about 27°C.[26] Spawning occurs in the spring, beginning during May in the southern extremes of the range and in June or July in the northern latitudes. Spawning activity is generally complete by August, though some spawning may occur during that month. It is possible to obtain off-season spawning by exposing channel catfish to an artificial winter (by cooling the water to 17°C or below for a period of time and then gradually warming the water to within the spawning temperature range).[99]

1. Broodstock Selection and Stocking

Channel catfish broodstock are often maintained in one or more communal ponds throughout most of the year and are stocked at densities of about 375 fish/ha[100] or sometimes higher in the spring. Typically, fish of from 0.9 to 4.5 kg are utilized as brooders[101] since larger fish are difficult to handle (Figure 5). Catfish may, under some conditions, become sexually mature during their second year of life, but most are in their third year before first spawning.

Channel catfish are most commonly spawned in open ponds, in pens or in aquaria, though raceways and tanks have also been used.[102] In the pond-spawning technique, adult fish are allowed to select their own mates from among the population within the pond. When the pen technique is employed, the culturist selects a male and female and places them inside an enclosure (pen) within the pond, thereby imposing some selection

FIGURE 6.   Catfish egg masses adhere to spawning nests and may be removed to indoor hatching facilities by the culturist.

on the spawning process. Similarly, in the aquarium spawning technique, the culturist places a pair of adult fish in a suitable-sized aquarium, injects them with hormones and induces them to spawn.[103] The aquarium technique was first employed as a means of studying the spawning behavior of channel catfish and is not often utilized by commercial culturists.

The augmentation of existing broodstock numbers or replacement thereof involves some selection on the part of the culturist. If only the largest fish are selected from among a group of catfish spawned during the same season, the likelihood will be that the selected fish will be dominated by, or even be composed completely of males[104,105] since males grow more rapidly than females. Selection of broodstock should involve individuals sufficiently large to be sexed with a reasonable degree of accuracy. The sexes can be distinguished with relative ease, particularly when adults approach spawning condition.[98]

Studies on the best ratio of males to females in spawning ponds for channel catfish have been conducted.[106,107] Bondari[107] concluded that there was no difference in the efficiency of reproduction when males were stocked at ratios of 1:1, 1:2, 1:3, or 1:4 with females. That conclusion was based upon spawn weight, egg weight, and various hatching traits. Commercial producers often stock spawning ponds with one male for every two females.

2. Spawning and Egg Hatching

In ponds and pens, some type of nest must be provided, within which the catfish lay and guard their eggs. Various types of nests have been used,[1] with milk cans being among the most popular. Any enclosure into which the adults can swim and undergo

FIGURE 7.   After a period of 5 to 10 days incubation, the eggs hatch into sac fry which will begin to feed following yolk sac absorption.

spawning activity is suitable. A typical female will lay from 6600 to 8800 eggs per kilogram of body weight.[98]

Once a female has spawned, the male chases her from the nest and begins to guard the eggs, which are laid in a yellow, adhesive mass (Figure 6). In nature, the male will constantly fan the eggs with his fins to provide a current which keeps oxygenated water flowing through the egg mass. Many culturists collect egg masses from spawning ponds or pens at intervals of 24 to 96 hr and remove them to a hatchery.

In the hatchery, the eggs are generally placed in a trough of flowing water which is well aerated. Paddle wheels within the hatching trough help keep the water in motion to assist in aeration. The first paddlewheel hatching trough was powered by a water wheel,[108] though modern ones are operated by electric motors. Jar hatching can also be employed,[109] but is not as common as the trough technique.

The time required for hatching depends, in part, on ambient water temperature. Within the normal spawning temperature range of the species, channel catfish eggs can be expected to hatch in 5 to 10 days.[110] First feeding follows yolk sac absorption, which generally takes less than a week after hatching has occurred (Figure 7).

The development of catfish eggs has been addressed by various authors.[98,111,112] Egg diameter is positively correlated with fish growth during the first 30 days after hatching, but not thereafter.[113] It appears as though other factors (genetic and/or environmental) become more important than initial egg diameter after 30 days.

3. Fry Stocking

Catfish ponds are prepared in the spring to receive fry. Preparation usually includes fertilization to promote plankton blooms and may involve treatment of the pond surface with diesel fuel or a mixture of cottonseed oil and diesel fuel (1:4) to control aquatic insects.[114] The oil floats on the surface of the pond and impairs respiration of insects which surface to obtain oxygen.

While some culturists allow fry to hatch within spawning ponds, most remove the eggs to hatcheries and stock fry following hatching and the determination of fish numbers (generally by volumetric displacement estimation). Many culturists provide finely ground feed to swimup fry and maintain the young fish in the hatchery until they are accustomed to feeding on prepared diets. Normal fry stocking densities are between 125,000 and 625,000/ha.[101] The fish are maintained at high density until they reach fingerling size (a few cm in length), after which they are transferred to production ponds, usually during the growing season subsequent to that in which they are hatched.

B. Genetics

Examination of large numbers of channel catfish will reveal to even the most careful observer that the range of traits from which to select for genetic improvement is very narrow. Thus, many producers select from among their most rapidly growing fish for broodstock replacement. Yet, in every sib group of catfish, there are those which grow rapidly, those which are intermediate in growth rate, and those which grow slowly. That relationship does not seem to become skewed toward the more rapidly growing fish as a result of selection. There is some reason to believe that selection for rapidly growing broodstock may, in fact, be selection for aggressiveness, since fish which are able to compete effectively for food are generally those which grow the most rapidly. This is borne out by the fact that if the larger fish within a population are removed, the smaller ones will become more competitive for food resources and their growth will improve markedly.

While size at a given age is most often utilized as the characteristic upon which selection is based, it has been demonstrated that body size is not strictly under genetic control,[115] so other traits have also been evaluated.[116] Bondari[117] indicated that a combination of within and between family selections should be effective in improving the genetic characteristics of channel catfish with respect to body weight and total length. He also indicated that female fish which show early growth inferiority should not be employed as broodstock.

It is likely that most of the catfish broodstock presently being utilized in the U.S. represent a relatively small gene pool which can be traced back to Federal hatcheries. For that reason, comparison of groups of progeny obtained from various hatcheries may not reveal distinct genetic differences since all the fish may have come from a common original source. In order to overcome this problem investigators have attempted to obtain broodstock from isolated natural populations or have attempted to improve performance through crossbreeding the various catfish species native to the U.S. Such attempts at genetic improvement have not, however, always been successful. For example, Broussard and Stickney[118] compared two domestic and two wild strains of channel catfish and found that higher growth rates occurred in the domestic strains, though no significant differences in survival were obtained among the strains.

Geographic region of origin as well as species may play a role in catfish performance. In at least one study, channel catfish from northern regions outperformed their southern conspecifics during winter though white catfish outperformed not only channel but also blue catfish.[119] Hybrids from blue x channel crosses have been shown to grow more rapidly than either parent,[120,121] were easier to seine, more uniform in size, and yielded a better dressout percentage.[121]

Spawning success of hybrid blue x channel catfish has been demonstrated to be much poorer than that of either parental group.[122] Further, blue catfish female x channel catfish male offspring cannot be produced as easily as the reciprocal cross. The blue male x channel female hybrid has been shown to grow more rapidly than the more difficult cross and has other desirable characteristics from a culture standpoint.[123] Regardless, few commercial producers utilize hybrid catfish in their operations at the present time. The industry continues to be dominated by channel catfish.

The heritability of various traits in channel catfish appears to be most heavily influenced by the male of the species.[124] Among the male-dominated traits are swim bladder shape, anal fin ray number, external appearance, growth rate, morphometric uniformity, and susceptibility to capture by seining. The degree to which the male dominance factor can be used to improve commercial catfish stocks remains to be demonstrated.

Induction of polyploidy in catfish has been utilized in attempts to improve performance. Wolters et al.[125] induced triploidy by cold-shocking fertilized eggs at 5°C for an hour beginning 5 min after fertilization. Triploid fish were found to be significantly heavier than diploids at 8 months of age and older.[126] Triploid fish were also found to convert feed more efficiently than diploids, though there was some alteration in gonad histology among both sexes of triploid fish as compared to diploids.

Albinism has been common in some commercial catfish hatcheries and is rare in others. The trait is generally thought to be inherited in a manner which follows that of other species, that is, albinism is the result of a homozygous recessive gene. However, the incidence of albinism in channel catfish has been shown to increase with exposure to various heavy metals,[127] developing as a result of exposure of either eggs or adult fish. Metals which have led to increased albinism are aresenic, cadmium, copper, mercury, selenium, and zinc, all at concentrations ranging from 0.5 to 250 mg/ℓ.

VIII. FEEDING AND NUTRITION

Feed represents the highest variable cost associated with commercial catfish farming, accounting for as much as 55% of annual operating expense.[128] Large farms may require feed inputs on several tons per day, particularly late in the growing season. Modern catfish feeds are made up of relatively few ingredients, though price can vary considerably on a temporal basis because of fluctuations in the cost of the feedstuffs and energy which are utilized in feed manufacture.

Feeding practices for channel catfish have been fairly well standardized and the nutritional requirements of the species established.[1,21,129 131] Research continues in both areas, however, since even small improvements in feed quality or in the manner in which feed is presented can have considerable impact on the economics of production.

A. Types of Catfish Feed

Prepared feeds for catfish are of two basic types. A typical feed may be prepared by pressure pelleting or by passage of the mixed ingredients through an extruder. The major difference between the two techniques involves the amount of pressure and heat to which the feed mixture is exposed. Pressure pellets are exposed to far less heat and pressure than extruded products; thus the former type tends to retain heat-labile vitamins much better than the latter. Pressure pelleted feeds sink when introduced into water, while extruded pellets may either sink or float depending upon the type of process involved. When certain formulations exit from the extruder, the carbohydrates in the diet become almost instantaneously gelatinized, a process which traps air within the pellets and causes them to expand. Such pellets tend to float in water, often for periods in excess of 24 hr.

Because the amount of energy involved in the manufacture of extruded pellets is significantly greater than that employed in the pressure pelleting process, extruded diets tend to be the more expensive of the two. Another difference in cost relates to the fact that extruded feeds must be overfortified with heat-labile vitamins in order for the final feed to meet the requirements of the fish.

The major advantage of utilizing expanded diets is the flotation which enables the fish farmer to observe feeding activity. If the fish go off feed it may be a sign of water quality depletion or disease. Most commercial catfish farmers in Mississippi utilize floating feed, distributing it with feeding devices pulled behind tractors.[29]

Both pressure-pelleted feeds and expanded pellets can be ground to small sizes for use in feeding fry and early fingerling stages of channel catfish. It is generally acknowledged that the formulation utilized for young fish should not be the same as that utilized for older fish. In general, young rapidly growing fish require higher levels of nutrients and energy than do older fish. Diets specifically formulated for fry and early fingerling channel catfish have been formulated,[132] though many producers utilize trout starters or finely ground regular catfish rations for their young fish.

B. Feeding Regimes

Within the optimum temperature range of the species, channel catfish less than 1.5 g should be fed at 3-hr intervals throughout the day when maintained under conditions where natural food is not readily available. Once the fish are larger than 1.5 g, feeding frequency can be reduced to four times per day.[133] Fry catfish are most commonly fed to excess (as much as 50% of body weight daily), but once they reach about 0.25 g, a feeding rate of 10% of body weight per day can be established.[133] That can be further reduced to about 5% when the fish reach 4 g under intensive culture conditions.

Most commercial producers feed pond-reared catfish *ad libitum;* that usually works out to between 3 and 4% of body weight per day. Studies have shown that best growth occurs when catfish are fed to satiation twice daily[134] as compared with higher or lower feeding frequencies. *Ad libitum* feeding will produce better growth of channel catfish than rates equivalent to 50, 75, and 110% of the *ad libitum* rate.[135] When feed is restricted, e.g., to 2% of body weight daily, improved food conversion efficiency may occur when the fish are fed more often than twice daily.[136]

It was once felt that feed should be limited when the fish reach sufficient size and/or density that satiation or percentage body weight feeding requires more than a few tens of kilograms per hectare per day (usually less than 50 kg/ha/day). In recent years, some Mississippi catfish farmers have been feeding at rates in excess of 100 kg/ha/day, particularly late in the growing season. In a study which compared water quality at feeding rates which reached 34, 56, and 78 kg/ha/day,[137] no dissolved oxygen problems occurred at the lowest feeding rate, but dawn DO was typically below 2.0 mg/ℓ in ponds fed at the highest rate. Mortality was higher in the ponds fed at the highest rate and net profit was considerably reduced as compared with the low and intermediate rates of feeding. Best profit was obtained from ponds fed a maximum of 56 kg/ha/day.

As environmental temperature increases, so does the metabolic rate of catfish, at least up to a point. Widely used recommendations for feeding rate as a function of temperature have been developed[138] (Table 2). In climates where there are no extended periods of ice cover and at least some warm days occur during the winter, maintenance of pre-winter body weight, and even some growth can be obtained by feeding on warm days or every other day.[139] Studies with demand feeders have shown that spring onset of feeding did not occur until the temperature reached 12°C and that fall feeding activity stopped at temperatures below about 22°C.[140] When the fish do not have to trigger a demand feeder, they will ingest feed at considerably lower temperatures. Some reduction in winter feed costs may be realized by employing a lower protein feed than

Table 2
RECOMMENDED FEEDING
RATES FOR CHANNEL
CATFISH AS A FUNCTION
OF ENVIRONMENTAL
TEMPERATURE

Temperature (°C)	Feeding rate (percentage of fish body weight)
>32	1
21—32	3
16—21	2
7—16	1
<7	None

From Bardach, J. E., Ryther, J. H., and McLarney, W. O., *Aquaculture,* Interscience, New York, 1972, 1. With permission.

that used during the growing season. In some cases, a 25% protein diet may perform as well as a 35% protein diet during winter.[141]

Distinct feeding behavior patterns have been observed in catfish ponds.[142] During much of the time, the fish occupy home areas, but they make daily trips to the area where feed is made available, taking specific routes from the home area to the feeding area. Home areas tend to be in shallow water during the spring and fall and in deeper water during the summer and winter. When demand feeders are used, the larger fish feed first.[141] A similar pattern may occur when feed is provided once or twice daily by the farmer, particularly if the fish are fed in a restricted portion of the pond.

Many fish farmers feed only 6 days a week, and for some, a 5 day a week feeding regimen is employed. Studies have demonstrated, however, that even deprivation of feed for a single day will cause the fish to alter their feed intake.[143] For each day off feed, it will take an additional day for a fish to return to its original feed intake level. In addition, each day of feed deprivation requires 2 days for recovery of original growth rate.

C. Nutritional Requirements

Animal feeds are formulated to contain three energy-bearing components: protein, carbohydrate, and lipid, as well as vitamins and minerals and an appropriate amount of energy. Proper balance and levels of each of the components included in the diet of a particular species must be developed through research. With respect to channel catfish, the nutritional requirements have been under investigation for several years and diets which perform satisfactorily have been developed.[21]

1. Protein

The most expensive component of a catfish diet is protein. It must be provided not only in the proper quantity, but must contain the ten essential amino acids required by catfish (arginine, histidine, isoleucine, leucine, lysine, methionine, phenylalanine, threonine, tryptophan, and valine[129]) in the proper amounts. In addition, protein must be properly balanced with energy if it is to be efficiently utilized. A major goal of the fish nutritionist is to maximize the use of dietary protein in the development of new animal tissue (growth) and minimize its use as an energy source by providing the proper levels of carbohydrate and lipid for the latter use.

Modern complete channel catfish rations for fingerling fish in growout facilities generally contain about 32% crude protein. Fry and early fingerlings require higher levels of protein, perhaps 50% at first feeding. The protein level can be reduced over the first several weeks of feeding until the 32% level is reached.[132] Dietary protein should be supplied, at least in part, by some type of animal by-product. Fish meal, meat and bone meal, and poultry meal are examples, with fish meal being the most widely utilized. More exotic animal proteins, such as shrimp by-product meal[144] could also be used if available in sufficient quantity at a competitive price. In general, the contribution of animal protein is kept as low as possible, but is sufficiently high to ensure that the level of the first limiting amino acid (lysine[145]) is met. From a practical standpoint, meeting the lysine requirement will bring the other essential amino acids into the formulation at or above their minimum required levels.

Animal protein is required in channel catfish diets formulated to provide for all of the nutrient requirements of the species. Satisfactory growth may be obtained in ponds from diets containing only plant proteins,[146] particularly when stocking density is low enough so that some natural food is available. For complete diets, animal protein may be required at dietary levels of about 10% in order to supply the proper amino acid minimums, provided suitable plant products make up the remainder of the protein component of the diet. Among the most widely utilized plant proteins are soybean meal and corn meal. Wheat is sometimes utilized as a binder and provides some crude protein to the diet. Other plant proteins which have received some use, particularly during periods when their prices are sufficiently low to bring them into a feed formulation on a least-cost basis, are peanut meal and cottonseed meal. The latter typically contains a toxic chemical known as gossypol. Growth of channel catfish may be inhibited by dietary levels of cottonseed meal above 17.4% or by diets containing a free gossypol level above 0.09%.[147]

While supplies of the required dietary proteins, particularly plant proteins, are readily available in the U.S., there has been some interest in evaluating untraditional plants as potential sources of protein in catfish diets. Included have been various aquatic plants[148] such as duckweed and water hyacinths.[149,150] Because of the high percentage of water in such plants (approaching 95% of total biomass), their use would have to be confined to the immediate vicinity of harvest or some type of drying facility would have to be established at the harvest site since transportation of the wet plants would be prohibitive.

2. Carbohydrate

Energy is obtained from carbohydrates, but sugars and starches are stored only in small concentrations within the bodies of fishes. Relatively high levels of carbohydrate (up to about 40%), often primarily in the form of starch, can be found in channel catfish rations. The presence of starch does not seem to affect protein digestibility, but dietary cellulose does reduce starch digestion.[151] Diets containing high levels of fiber do not appear to provide added nutritional benefit to catfish,[152] and nutritionists generally consider fiber to have no food value for fish.[21,131] Cellulase enzyme activity (apparently acquired from intestinal microflora) has been found in channel catfish,[153] but the amount of cellulose digested in passage through the gut of the fish can be considered negligible.

Carbohydrate can be utilized in catfish diets to spare protein. That is, the carbohydrate can be utilized for various metabolic processes as an energy source, allowing dietary protein to be deposited in new tissues. Lipid can also be used in protein sparing.

Many of the dietary ingredients utilized as plant protein sources also contain high levels of carbohydrate. Examples are soybeans, corn, and wheat. Thus, when plant meals are added to fish diets, a significant portion of the energy they provide is in the form of carbohydrate, largely starch. Nutritionists generally formulate to balance protein and amino acid levels, so the carbohydrate present in dietary ingredients is not given much consideration.

3. Lipid

Dietary fats fall into several biochemical categories which can be collectively referred to as lipids. On a per unit weight basis, there is approximately twice the number of calories in a lipid than are found in either a protein or carbohydrate. Thus, lipid is an excellent energy source. In addition, many dietary lipid sources are relatively inexpensive, readily digestible, and may contain fatty acids essential for proper growth and metabolism.

When practical dietary lipids such as beef tallow, oilseed oils, or fish oil are fed to channel catfish at environmental temperatures below about 25°C, there does not appear to be much effect of lipid quality or quantity on growth.[154,155] However, at temperatures within the optimum growth range of the species, diets containing beef tallow and fish oil outperform such oilseed products as safflower and soybean oils[156] and dietary lipid quantity may become more important.[157]

The reason for the better performance of channel catfish on diets containing beef tallow and fish oil, two lipids which are quite different chemically, remains to be completely explained, but a recent study[158] provides some support for the hypothesis that the high level of linolenic acid in oilseed oils has a depressive effect on growth. Fish oil has high levels of linolenic acid family fatty acids, but does not contain much of the acid for which the family was named. Beef tallow contains very little linolenic acid and is nearly devoid of other fatty acids within the linolenic acid family. More work is needed in the area of lipid metabolism research with channel catfish to determine if growth response is related to the absolute levels of dietary fatty acids or to the proportions of the various fatty acids in the diet.

4. Vitamins

Vitamins have been defined as organic compounds which are required by at least some species in small quantities for normal growth and health.[159] Included among those required by channel catfish are both the water-soluble (B complex and C) and fat-soluble vitamins (A, D, E, and K). Vitamin packages are routinely added to practical catfish feeds, particularly when the feed is formulated for fish stocked in situations where natural food is unavailable or present only in low quantity. The levels of each vitamin required for proper growth and health have been outlined and widely reported, though some work is continuing in the area.[21,129-131] Vitamin D was at one time thought to be nonessential for channel catfish, but more careful study has revealed that deficiency symptoms can be induced in the absence of the vitamin.[160] Reduced weight gain, poor food conversion efficiency, and an effect on bone mineralization have been found in association with extremely low levels of dietary vitamin D.[160,161] Similarly, it has been difficult to demonstrate a clear requirement for vitamin K.[162]

Various of the vitamins are subject to loss from diets through exposure to heat, moisture, or light. Among the most labile of the vitamins is ascorbic acid or vitamin C, which is quickly destroyed by heat. To ensure that the level of vitamin C in finished rations is sufficient, most formulations call for at least a fivefold increase in the formulation over the requirement of the fish.

Deficiency signs in channel catfish are clearly indicative of a particular vitamin shortage in some instances and are somewhat vague in others.[1,21] Vitamin C is one for which deficiency signs are both clear and somewhat unique. Caged channel catfish will show such signs as scoliosis, lordosis, and depigmentation of the skin on the back within 12 weeks when stocked at 400 fish/m³ and fed diets deficient in the vitamin.[163] Broken backs may result from weakness in the vertebral column caused by disruption of collagen-forming ability. Addition of vitamin C to the diet will lead to recovery in many of the affected fish, but fish with fractures of the vertebral column should be destroyed. All fish stocked under intensive culture conditions should receive only feed which is well fortified with vitamin C. Megalevels of the vitamin (up to 50 times the normal level) seem to impart some resistance in channel catfish to certain bacterial infections.[164]

5. Minerals

Most of the minerals required by channel catfish can be obtained from the water and are present in the natural feed ingredients used in the formulation of practical diets. In research diets of a semipurified or purified nature, mineral packages are added which contain the major minerals (calcium, phosphorus, sulfur, sodium, chlorine as chloride ion, potassium, and magnesium) and the trace minerals (iron, copper, iodine, manganese, cobalt, zinc, molybdenum, selenium, and fluorine). In practical diets, most of the major and trace minerals will be present in sufficient quantity within the natural feed ingredients, though calcium and phosphorus are often added to ensure that the diet contains the recommended level for those minerals.[165,166]

The availability of minerals can vary considerably depending upon the form in which the mineral is fed. For example, phosphorus is much more readily available from monosodium or monocalcium phosphate than from dicalcium phosphate, fish meal, soybean meal, or cereal products.[167]

Sodium chloride does not seem to affect catfish growth when fed at levels of up to 2% of the diet.[167] The elaboration of deficiency signs for other minerals may also be difficult to elicit. For example, no deficiency signs were discerned from channel catfish fingerlings reared on diets containing various levels of copper from 0 to 32 mg/kg of feed with a background level of 1.5 mg/kg which could not be eliminated from the diet.[168]

IX. DISEASE

During the mid-1960s, only about 5% of the fish farmers in the U.S. reported serious problems with disease outbreaks.[169] That figure may still be generally valid throughout much of the industry, but in Mississippi, about one third of the catfish farmers reported severe disease epizootics in a survey published in 1980.[29] The differences may relate to the significant increase in stocking density which has occurred during the years between the surveys or to a number of other reasons, but in any event, disease does play a role, and sometimes a major one, in catfish production.

For purposes of our discussion, the term disease includes maladies caused by such diverse organisms as viruses, bacteria, fungi, and a variety of parasites. No attempt has been made to outline all of the diseases to which catfish are subject, but reviews on that subject are available.[1,170] Our intent is also not to provide detailed information on disease treatment. Instead, emphasis has been placed on some recently published information relative to channel catfish disease problems.

One of the major problems facing fish culturists is a paucity of treatment chemicals which have been cleared for use on food fish in the U.S. Meyer et al.[171] indicated that only acetic acid, sodium chloride, sulfamerazine, and oxytetracycline (Terramycin) were cleared for foodfish use as of 1976. Since then formalin has also been cleared by the federal government. The need for additional therapeutic agents for the treatment of fish diseases remains critical, and the recent decision by the U.S. Food and Drug Administration to include fish as a member of the "minor species" category may mean that the range of treatment chemicals available to fish farmers will increase in the future.

Bacteria of the genus *Edwardsiella* have been known for over a decade to be pathogenic to channel catfish,[172] with *E. tarda* being the first species described. That bacterium causes gas-filled, foul smelling lesions to form in the muscle tissue of the fish. Control was affected with oxytetracycline in the feed. Treatment with that antibiotic may be in jeopardy, however, since the discovery of a strain of the bacterium which is resistant to Terramycin.[172]

In 1979, a new bacterial disease of channel catfish was described[174] which was attributed to a different species of *Edwardsiella* that could also be controlled with oxytetracycline. The new species, now known as *E. ictaluri*,[175] causes a disease known as enteric septicemia which occurs at temperatures from 25 to 30°C.[174] Clinical signs are not consistent, but may include such things as fish which hang listlessly at the water surface with their tails down, those which display rapid swimming in circles, and those which swim in a spiral fashion. Hemorrhages may occur around the throat and mouth, the gills may be pale, and the fish may show signs of exophthalmia. A common sign is the occurrence of an open lesion in the frontal bone of the skull. Gross internal pathology may include hypertrophy of the kidney and spleen, areas of hemorrhage and necrosis in the liver, bloody fluid in the peritoneal cavity, and hemorrhaging in the adipose tissue, inner walls of the visceral cavity, dorsal musculature, and intestine. In small fish, these signs can be easily confused with those of channel catfish virus disease.

Channel catfish are not the only species susceptible to *E. ictaluri*. *Tilapia aurea* may also contract the disease,[176] though golden shiners, largemouth bass, and bighead carp appear to be immune.

Work has progressed in recent years with the hyperosmotic infiltration technique of vaccinating fish against disease. The technique involves placing fish in a hypoosmotic solution and allowing them to dehydrate slightly, then placing them in a hyperosmotic solution containing the vaccine. The vaccine infiltrates the fish during rehydration. In a study comparing hyperosmotic filtration to injection for development of immunity to the bacterium *Aeromonas hydrophila*,[177] it was found that injection was the better technique. However, from a practical standpoint, the hyperosmotic infiltration method is superior. Problems involved with injecting several thousand or hundreds of thousands of individual fish are not difficult to appreciate.

An intriguing problem occurred during the winter of 1981 in Mississippi, being called the "winter kill".[178] It was apparent primarily in intensively managed ponds and led to a steady daily mortality of fish within affected ponds during the winter period. Various attempts at treatment did not appear to affect the problem. Clinical signs include a rough appearance of the fish with loss of mucus and the development of dark areas on the sides or back. That may be followed by loss of the external skin layer leading to the appearance of white areas on the skin which are subsequently invaded by fungus. Other signs, though not always present, include sunken eyes, gill lesions, and in some cases, skin lesions. A variety of bacteria have been isolated from the skin lesions, with none of them common to all fish.

The winter kill problem was most pronounced in ponds with fish of large size. Ponds stocked at densities of less than about 10,000 fish/ha did not generally exhibit the problem. There was some indication that high levels of ammonia influenced the problem.

Recommendations for avoiding the problem included reduction of pond biomass, maintenance of good water quality, and getting the fish on sinking feed in the fall in an attempt to keep them feeding in order to help maintain their physiological condition during winter.[178] Of some interest is the fact that the problem occurred during a year when many producers maintained unusually high densities of fish over winter. During the winter of 1982, when fewer fish were maintained during the winter months, there was no recurrence of the disease.

X. HARVESTING AND PROCESSING

Once the fish have reached market size, they must be captured, transported to the processing plant, and made into products to be sold in various retail outlets. The responsibility of the catfish culturist does not end until the fish reaches the table of the consumer since production of a substandard fish will almost surely lead to loss of a customer. Many fish culturists have not been too concerned with quality control in the past, but with the occurrence of significant off-flavor problems throughout the commercial catfish industry in recent years, both producers and processors have become acutely aware of the need to deliver a high quality product.

A. Harvesting

The harvest of channel catfish from tanks, raceways, and cages is a relatively simple matter. Water volume can be sufficiently reduced in those types of intensive culture chambers to allow for dipnetting of the fish. In ponds, the most widely used type of culture chamber, seining is the standard technique.

A typical seine is at least a few tens of centimeters deeper than the maximum depth of the pond to be harvested, has a float line at the surface which is buoyed by plastic or cork floats, a mud line at the bottom which is comprised of several strands of rope to help keep the net on the bottom, and a bag in the middle into which the fish are crowded during the seining operation. A seine should be about 1.5 times longer than the width of the pond.

In the early years of the industry it was common practice to harvest all the fish in a given pond simultaneously. While this practice is still employed in some areas and by some individuals throughout the industry, common practice today is to partially harvest each pond at intervals of a few weeks or months and to restock fingerlings periodically, replacing fish which have been marketed.

When the entire pond is harvested, an initial seine haul might be made after which the water depth is reduced by about 50% and then another seine haul is made. Subsequent dropping of the water level followed by seining is done until the entire population has been recovered.

If the partial harvest technique is used, the water level is generally maintained throughout the seining operation. Large mesh seines are used to select for only fish of market size; most of the submarketable fish will escape through the net, and those which don't are returned to the pond. Partial harvest can be employed for extended periods, but complete harvest will be required, accompanied by total pond drainage at least every few years. When total harvest is accomplished stunted fish can be removed from the population and the pond can be allowed to dry, providing an opportunity for accumulated organic matter to be oxidized. The pond bottom may be disced and treated with lime at the same time. Careful monitoring of water quality and production will provide the catfish farmer with an indication of when complete harvest should be employed in a given pond.

Large ponds may require seines of over 100-m length. Such seines cannot be pulled with manpower alone. Instead, trucks or tractors are employed.

Once the fish have been gathered, a live car may be attached to the bag end of the seine.[179] The live car forms a net pen which serves as a corral for the fish, allowing the seining crew to move to another pond.

Catfish are virtually always hauled to market alive in trucks of various sizes equipped with insulated holding tanks and provided with supplemental aeration. Agitators, blowers, compressed gas cylinders (air or oxygen), liquid oxygen tanks or some combination of these may be available on the hauling truck. It is important that the fish reach the processing plant in good condition or the shipment may be refused. Large farmers may purchase their own hauling trucks, though custom live-haulers are available in regions where the demand for their services is high.

B. Processing

The large commercial catfish processing plants in the U.S. are, not surprisingly, located for the most part in the heart of the major catfish-producing areas, predominantly in Mississippi. Texas is an example of a state which has no commercial processing plants; processing is left up to the individual farmer in most cases.

Processing techniques are fairly well standardized,[180] though there has been a move toward increased automation in some of the newer plants. In the past, virtually all of the processing was done by hand labor.

Various types of catfish products are being marketed by catfish processors, with common forms of the product being filets, steaks, and skinned, gutted, and deheaded whole fish. The products may be rapidly frozen in a blast freezer or marketed chilled on ice. Farm-raised catfish can be found throughout most of the U.S., particularly in the larger cities. A considerable amount of effort has gone into marketing in recent years as the industry has expanded. The phobia against catfish which once prevailed in states above the Mason-Dixon line, while not completely overcome, has been significantly reduced. Quality control has played a major role in the marketing successes to date.

Off-flavors have been a major problem with pond-reared channel catfish for several years. The problem is not a new one, nor is it associated only with catfish;[181,182] however, it seems to occur more frequently in catfish than in most other cultured species. It has become increasingly a problem as the level of culture intensity has increased. A recent study showed that off-flavor tended to intensify with increased feeding rate.[183]

The off-flavor problem is commonly described as an earthy-musty flavor and odor, though flavors described as sewage, stale, rancid, metallic, moldy, weedy, and petroleum-like have also been identified by taste panelists.[184] The primary chemical responsible for off-flavor appears to be geosmin (trans-1,10-dimethyl-trans-9-decalol) which has been isolated from a number of microorganisms, particularly blue-green algae.[185-188] Transfer of affected fish to geosmin-free water (for example, placing the fish in a tank which receives well water) will lead to disappearance of the off-flavor problem within a few days.[188-190]

It is common practice at commercial processing plants in Mississippi to conduct as many as three taste tests on representative fish from a single pond before the fish are accepted by the processor. The first taste test may be conducted 2 weeks before the scheduled date of harvest, the second 3 days before harvest, and the third at the time the truckload of fish reaches the processing plant. If off-flavor is detected, the fish will be rejected until they can pass the taste test. Preharvest testing is highly desirable since once loaded on a hauling truck it is difficult to return fish to a pond without suffering heavy loss. Most farmers do not have the facilities available to purge their fish in tanks with well water and depend upon live-hauling directly from the pond to the processor.

If a pond fails the test, it may be several weeks or months before the problem is corrected. Under commercial conditions, the farmer typically must hold the fish within the affected pond until the algae which caused the problem die back and the geosmin or other off-flavor-causing chemical is metabolized.

XI. FUTURE

As indicated previously, the channel catfish industry has reached the $100 million/ year level in terms of returns to the commercial producers in the U.S. Continued expansion will depend to a large extent upon further market development, particularly with respect to increased acceptance in areas which have not traditionally consumed catfish. If market expansion continues, production increases must accompany increased production per unit area, expansion of facilities, or a combination of the two. two.

Facility expansion is possible in some regions of the U.S., but is severely limited in areas where high land prices, limited water supplies, and improper soil conditions prohibit the economical development of new catfish farms. While little can be done about existing soil type, land prices and the availability of suitable quality water are constraints which can be expected to intensify in the coming years.

Present production in existing facilities may be nearly maximum under the management schemes currently in use. Further increases would seem questionable in light of the deterioration in water quality that is currently affecting many producers. It appears as though present pond culture practices are pushing production to its limit. Modern channel catfish farmers would seem to be operating at the edge of disaster throughout much of the growing season, so monitoring of the culture situation has become critical. What is now needed is the predictive capability to appropriately alter management as a means of avoiding an imminent problem.

Predictive capability will come to the catfish farmer, and to aquaculture in general, as computer simulation models of fish farming operations are developed and improved. Some early attempts at predicting pond production through computer simulation suggest that the technique holds considerable promise,[191] but a significant amount of additional experimental data will be required before a high level of confidence can be placed in computer models as decision-making tools for commercial fish farmers. Perhaps the way in which computer modeling currently makes its greatest contribution is by suggesting areas where additional research is needed. Once conducted, the results of such research can be fed back into the model and utilized to improve the quality of output obtained.

Chapter 3

CARP AND BUFFALO

Robert B. McGeachin

TABLE OF CONTENTS

I. INTRODUCTION AND HISTORY OF CARP CULTURE

The common carp *Cyprinus carpio* (Figure 1), is a hardy species, tolerant of handling and poor water quality conditions, which is widely cultured around the world, though not popular in the U.S. It is especially tolerant of low-dissolved oxygen concentrations and has the capacity for anaerobic respiration over short periods.[192,193] The common carp is omnivorous and readily accepts prepared diets. It grows relatively rapidly in warm water, can be easily spawned in captivity, and has high fecundity. All of these traits contribute to the excellence of the species for culture. The problem of intercillary bones can be mitigated by various cooking methods, and off-flavors can be eliminated by feeding commercial diets and maintaining good culture conditions.

According to Balon[194] the species is native to central Asia. There are three probable subspecies: (1) the wild rheophilic ancestral form, (2) the European carp, *C. carpio carpio*, and (3) the "big belly" eastern Asian carp, *C. carpio haematopterus*. Since the fish are geologically young — no older than the Pleistocene — there is little difference among the subspecies. Within domesticated populations, however, there are a number of strains which have been highly selected and which appear and perform differently.

The common carp was one of the first fishes to be cultured and domesticated. A Chinese treatise on fish culture by Fan Li, written in the 5th century B.C., provided recommendations for carp culture.[195] Common carp continues to be widely cultured in China, Japan, and Southeast Asia.

A second, independent effort at domestication began in Europe during the 1st century A.D.[194] Wild carp from the Danube region were captured and shipped live (probably in wet moss) to Rome where they were maintained in special holding ponds called "piscinae" until eaten. This practice continued into the 6th century with the Visigoths and spread north with monks during the development of monasteries. The keeping of fish in ponds was especially important for monks since the fish could be consumed during the many meatless fasting days which were observed. Eventually, breeding was conducted in holding ponds and by the late Middle Ages carp culture had spread throughout eastern and central Europe.

The common carp was first brought to North America in 1831[194] and the first domestic carp were successfully introduced by Rudolph Hessel in 1877 for the newly created U.S. Fish Commission.[196,197] The 227 leather and mirror carp and 118 scaled carp in the shipment were temporarily kept in Baltimore and then moved to Washington, D.C. where they were bred in the reflecting pools of the Washington Monument. The U.S. Fish Commission started a program to rejuvenate the rapidly depleting inland fisheries of the U.S. by stocking carp in public waters and fish ponds on the request of private individuals. The intent was to provide inexpensive food in the face of depleted wildlife and fish stocks and rapidly rising livestock prices. Carp was particularly popular with and requested by European immigrants and their descendants who were familiar with the fish and its culture. The first carp were distributed in 1879 and by 1882 some 79,000 fish had been stocked in 298 of the 301 Congressional Districts.[196,197]

By the end of the carp stocking program in 1896, 2.2 million fingerlings had been distributed to about 50,000 individuals throughout the U.S. and about 250,000 fish had been stocked directly into public waters.[197] Individuals who received carp were instructed to keep the fish in clean water which contained a wholesome aquatic food supply and to supplementally provide good quality feeds. In practice, many persons stocked the fish in poor quality waters, fed them refuse and offal, or left them to scavenge. The result was off-flavor which helped the species fall into disrepute.

The grass carp or white amur, *Ctenopharyngodon idella* (Figure 2), is another species which has been introduced into the U.S. Native to the Amur River system on the border between The People's Republic of China and Siberia,[26] the species now inhabits

FIGURE 1. The mirror carp form of the common carp is characterized by having only a small number of scales, a trait which was developed through many generations of selective breeding.

FIGURE 2. Grass carp were introduced into the U.S. as an aquatic weed control agent.

Table 1
COMMON CARP PRODUCTION
(METRIC TONS) BY COUNTRY IN
1980

Country	Production
U.S.S.R.	118,756
Indonesia	43,318
Japan	33,524
Hungary	23,812
Romania	20,316
Yugoslavia	13,354
Czechoslovakia	12,476
U.S.	10,710
Bulgaria	10,139
Poland	8,793
Turkey	8,785
Israel	7,196
Democratic Republic of Germany	6,856
Italy	3,234
Austria	1,200

From FAO, Yearbook of Fishery Statistics Catches
and Landings, Food and Agriculture Organization,
Rome, 1981, 1.

many of the large river systems of East and Southeast Asia.[198] The adults feed on aquatic macrophytes, a trait which led to the introduction of the species to the U.S. for weed control in the 1960s. Common and grass carp are among the species utilized in Chinese polyculture ponds, a style of culture which dates back at least several hundred years.

Other carp species are under culture around the world, but to date, interest in carp culture in the U.S. has been centered on the two species discussed above. Therefore, our discussion is limited to common carp and grass carp.

II. STATUS OF THE COMMERCIAL CARP INDUSTRY

The Food and Agriculture Organization of the United Nations (FAO) estimate for common carp production worldwide in 1980 was 314,645 metric tons.[199] By species, common carp ranked 40th out of about 750 fishes in terms of size of catch. Although the FAO statistics do not differentiate aquaculture production from capture fisheries, a large portion of the total can be attributed to fish farming activity. Table 1 lists common carp production data from 1980 from the major producing countries. The production from Israel, Japan, and the European countries was primarily from aquaculture.[26,200-204] Common carp represent a minor component (about 5% of the fish stocked) of the Chinese polyculture system which generally employs three to six species; however, there are no data available on carp production in the People's Republic of China, though the tonnage can be assumed considerable.[205,206]

In the U.S. common carp are obtained only from capture fisheries. The stocking effort of the U.S. Fish Commission led to a significant inland fishery, but the relatively low market price and great abundance of wild carp has left little room in the marketplace for cultured carp. Presently, there is no commercial food production of common carp in the U.S., though some carp are produced for the ornamental and aquarium markets as well as for trotline bait.[207]

The grass carp was first introduced to the U.S. in 1963 for evaluation as a biological control for aquatic vegetation by the Fish Farming Experimental Station of the U.S. Fish and Wildlife Service in Stuttgart, Ark.[208] The fish was found to be an effective tool in the control of aquatic macrophytes in fish culture ponds and has been widely employed at low densities in states where use of the fish is not prohibited by law. Since it is an exotic species, the grass carp has been prohibited in many states pending the outcome of research to determine potential environmental problems associated with its introduction.

Production of fingerling grass carp for weed control began in Arkansas during 1972[209] and continues to be centered in that state as well as in Missouri and Kansas. Some of the grass carp stocked for weed control in polyculture with such species as channel catfish are eventually harvested and marketed as food fish. In Arkansas during 1978, 19 metric tons of food size grass carp were produced and 18 tons of fingerlings were reared for stocking.[210]

The grass carp has been introduced to Israel — also primarily for its role in aquatic vegetation control — and small numbers have reached the human food market.[26] The fish has also been introduced to Japan, where a reproducing population has reportedly been established in the Tone River. Commercial production of grass carp has not as yet been initiated in Japan, however.[204] Finally, grass carp have been introduced to the Sudan[198] and Egypt on an experimental basis to control aquatic vegetation in the irrigation canal systems which have been developed in those countries.

III. CULTURE SYSTEMS

A. Ponds

Common carp are cultured primarily in earthen ponds. The degree of culture intensity ranges from highly extensive (low stocking densities with no supplemental feeding or fertilization) to relatively highly intensive (high density stocking, fertilization, provision of complete diets, mechanical aeration, management of phytoplankton blooms). Pond production in Europe is mostly extensive in nature. In Austria, Belgium, France, and the Federal Republic of Germany, pond production involves no feeding or fertilization. Production averages about 300 kg/ha.[26] In those countries about 3 years are required to produce carp of 1.0 kg.

Traditional carp culture in Czechoslovakia is also extensive, but fertilization with manure is practiced as a means of increasing natural food supplies. There is also some supplemental feeding with grain.[200] Recent emphasis has been to integrate carp culture with the production of ducks and geese. The birds provide fertilizer and spilled feed to the water, thus increasing carp production.

In Poland, pond culture of carp leads to production levels of around 500 kg/ha.[202] The addition of grain as supplemental feed has been shown to increase production to levels of 1000 to 1200 kg/ha. Experiments in which pelleted feeds have been provided led to further increases in production to 4000 kg/ha, while the use of continuous aeration doubled the latter production figure.

Carp pond production in Hungary averages about 1000 kg/ha.[26] An experimental system of 10-year crop rotation has been initiated which consists of 5 years of carp and duck culture in ponds followed by 2 years of alfalfa, then 3 years of rice.

Intensive carp culture is practiced in The People's Republic of China, Japan, and Israel.[26,201,204] Carp production in those countries is enhanced by the moderate climates which can be found in many regions; thus, carp production requires only 2 years.

In the People's Republic of China the stocking density of various carp species in polyculture is high and ponds are heavily fertilized with manure and agricultural wastes. The ponds are closely monitored for water color and clarity so dissolved oxygen problems can be anticipated and manuring rates adjusted accordingly. Average pond production ranges from 1500 kg/ha in northern China to 3750 kg/ha in the south.[205]

About half of the common carp produced in Japan are reared in ponds. Production ranges from 3700 to 5200 kg/ha.[26,204] Ponds are heavily stocked, complete diets are provided, mechanical aeration is available, and plankton blooms are carefully managed.

In Israel, carp production which was initiated by eastern European immigrants has gradually shifted over the past 50 years from extensive monoculture to intensive monoculture with the use of high protein pelleted feeds and mechanical aeration or intensive polyculture with daily manure applications and the feeding of supplemental grains.[26,211] Carp production has been as high as 3700 kg/ha in polyculture and up to 30,000 kg/ha in monoculture.[211]

B. Raceways

In Japan, 14 to 17% of carp production is accomplished in stone or concrete raceways.[26,204] The water flow at the most productive of the farms employing raceways ranges from 91 to 455 ℓ/sec and the carp are fed pelleted diets containing silkworm pupae. The yield from such systems is the equivalent to 2,203,000 kg/ha.[26]

C. Cages

Culture of common carp in floating cages in lakes and reservoirs accounts for from 28 to 37% of production in Japan.[26,204] Square cages, 9 m on a side and 2 m deep are constructed to float to a depth of 1.5 m. An additional meter of netting around the top of the cages help retain the fish which are subject to jumping. Pelleted feed with a protein content of 39% is provided four times daily. The fish are stocked at no more than 75m² and require 2 years to reach market size of 0.8 to 1.0 kg. Production averages are equivalent to 431,000 kg/ha.[26]

Experiments with carp fingerling rearing in cages utilizing the thermal effluent from a power plant to increase the length of the effective growing season have been conducted in Poland. The technique has led to a doubling in normal annual growth.[212] Cage culture has also been initiated in irrigation canals in Egypt.

D. Rivers, Lakes, and Canals

In the People's Republic of China many lakes, reservoirs, rivers, and canals have been developed into extensive fish culture facilities.[205,206] The inlets and outlets of lakes and reservoirs have been blocked off with nets, predators controlled, and carp stocked and supplementally fed. Coves are isolated with nets and used as nursery grounds for fry and fingerling carp which are later stocked into the main rearing areas. Total fish production ranges from 53 to 250 kg/ha.[206]

Sections of rivers and canals may be fenced off with flexible gates which allow the passage of boats while retaining fish. Predators are removed by fishing, and fry are stocked subsequent to rearing in adjacent pools on the banks of the rivers and canals. The waterways are fertilized with agricultural runoff from adjacent lands.

IV. STOCKING DENSITIES

A. Polyculture

Common carp and grass carp are often reared with other species of fish.[26,201,205,206,213-215] There are a number of variables that influence the species and stocking density used by polyculturists. These include:

1.   Economics of culture and market demand for the fish
2.   Ecological niche to be filled
3.   Natural food and prepared ration availability
4.   Water quality, availability of water, and aeration equipment
5.   Size of fish at stocking
6.   Size of fish desired at harvest
7.   Climate and length of growing season
8.   Energy and labor available for stocking, harvesting, and processing

Polyculture in Israel generally consists of a mixture of common carp, tilapia *(Tilapia aurea)*, mullet *(Mugil cephalus)*, and silver carp *(Hypophthalmichthys molitrix)*.[201] Stocking densities for common carp at some high yield farms ranges from 2500 to 6500 fish/ha. That species may account for from 32 to 59% of total production. In many instances 2 year classes of carp are stocked simultaneously.

In The People's Republic of China polyculture involves a mixture of common carp, grass carp, silver carp, big head carp *(Hypophthalmichthys nobilis)*, black carp *(Mylopharyngodon piceus)*, mud carp *(Cirrhina molitorella)*, goldfish *(Carassius auratus)* and Wuchan fish *(Megalobrama amblycephala)*.[205,206] In ponds with a total stocking density of 15,000 fish/ha, common carp comprise 3 to 5% of the total and grass carp, which are sometimes the major species cultured, comprise from 12 to 55%. In the stocking of rivers, common carp commonly represent 20% of the total and grass carp, 35%.

B. Monoculture

The extensive monoculture practiced in most of Europe involves the stocking of common carp at densities of 300 to 600 fish/ha.[26] Experiments aimed at intensifying the culture of carp employed 300 fish/ha in unfed, unfertilized ponds, 900 fish/ha in fertilized ponds, and 4000 fish/ha in ponds which received pelleted feeds.[202]

Monoculture ponds in Israel may range in density from 1000 to 14,750 carp/ha.[201] Research in Israel has shown that at densities of 4000 to 20,000 fish/ha carp growth is inverse to density.[216] Economic analysis of carp production showed that the most profitable stocking density for monoculture was 16,000 fish/ha.

The 81 m^2 floating cages used in Japan are initially stocked with 43 to 75 fingerlings carp/m^2. Final production yields of from 100 to 200 kg/m^2 are achieved.[26,204] Stocking densities in raceways vary with water flow which ranges from 91 to 840 ℓ/sec. Average stocking in raceways is from 3 to 11 kg/m^2 of 70- to 150-g fingerlings. Production averages from 100 to 200 kg/m^2. When stocked at 181 fish/m^2, yields of 220 kg/m^2 have been obtained.[26]

V. REPRODUCTION AND GENETICS

A. Natural Spawning

The natural spawning season for common carp is during the spring when water temperatures reach 18 to 24°C and remain fairly stable.[201,217] Common carp lay adhesive eggs on plant substrates, so spawning ponds are supplied with mats of conifer branches or plastic artificial substrates.[201] Spawning ponds range in size from a few square meters to several hectares. In smaller ponds, once egg laying is complete, spawning mats are covered with wet cloth to prevent egg desiccation and are transferred to fry nursery ponds. Larger ponds are usually utilized both for spawning and fry nursing. After the eggs hatch, the spawning mats are removed and the brood fish are seined from the pond with large mesh nets.

During early spring the sexes are segregated into separate ponds to prevent premature spawning. Females of 2 to 5 kg and slightly smaller males are utilized as brood stock. They are stocked into spawning ponds, which have been allowed to dry to eliminate disease organisms and are filled immediately prior to stocking, at a ratio of 1:2 or 2:3 females:males. Densities of 10 females per hectare or less are utilized.[201,206]

The final environmental spawning cue for carp is spring flooding. Transportation of ripe carp and their introduction into the newly filled ponds appears to mimic that cue and leads to stimulation of spawning. Spawning usually occurs the morning after the fish are stocked into the spawning ponds. If spawning has not occurred within a few days, fresh water is flowed through the pond as a stimulus.

Fry nursery ponds are also dried prior to use and are fertilized with 60 kg/ha superphosphate and 60 kg/ha ammonium sulfate when filled. Chicken manure is added at the rate of 100 kg/ha every 2 weeks to stimulate zooplankton production which is the initial food of carp fry.[201,218] Fry ponds are treated every 2 or 3 days with vegetable or mineral oil or insecticides to kill predatory insects until the fry are large enough to avoid predation.[201,218] Ponds are also commonly treated at 3-day intervals with 0.2 ppm Bromex to prevent the establishment of monogenetic trematodes.[201]

Fry are reared in the initial spawning or nursery ponds until they reach from 0.2 to 0.5 g. They are then harvested and counted to establish survival and then stocked into secondary ponds at known rates.

Secondary nursery ponds are employed for rearing carp fingerlings of 10 to 50 g by the end of the first growing season.[201] Fry ponds should be stocked at densities that will lead to final standing crops at harvest of not more than 1000 kg/ha. The secondary nursery ponds are fertilized and the fish may receive supplemental feedings of ground grains.

B. Induced Spawning

Techniques for the induction of spawning in common carp are well established. With respect to grass carp, spawning can only be achieved in the hatchery as the fish will not produce viable fry in ponds.[201,206,218-222] Ripe brood fish are brought into the hatchery where the females are placed in individual containers and two to five males are placed together. The females should be weighed to provide an estimate of initial fecundity, with the average being 120 eggs per gram of body weight.[223,224]

Ripe brood fish are induced to spawn by injection with either carp pituitary extract, human chorionic gonadotropin, or leuteinizing hormone-releasing hormone at normal spawning temperatures (18°C or higher).[201,219] At low temperatures (13 to 15°C), spawning can be induced with carp pituitary extract and 17 alpha-hydroxy-20 beta-dihydroprogesterone.[220] When carp pituitary extract is employed the fish receive two injections 8 hr apart for common carp and 24 hr apart for grass carp.[219] Males are injected once at the time the female receives her second injection. Spawning occurs 10 to 14 hr after the final injection depending upon water temperature. The total injection rate for females is about 3 mg of dry weight pituitary per kilogram of brood fish with 10 to 15% in the first injection and the remainder in the second.

When eggs are found to flow easily from the female or a few released eggs are observed in the holding container, she is netted, wrapped in a towel, and the lower portion of the abdomen is dried.[219] One person holds the female and strips the eggs into a round basin held by a second person (Figure 3). Milt from two males handled in a smiliar manner is added to the dry eggs and the mixture is stirred. Milt which has been obtained in advance and stored under refrigeration may also be successfully utilized.[225] In the case of grass carp, water is added to the eggs after about 1 min of stirring to harden them.[219]

Incubation of the normally adhesive eggs of common carp in hatching containers requires that the eggs be treated with a salt and urea,[201,219,222] salt and talc, or salt and milk solution[226] to keep them from sticking. Egg volume for each spawn is recorded so

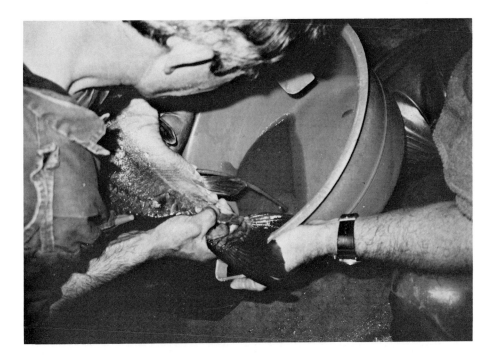

FIGURE 3.   Eggs being stripped from a common carp female at the fish hatchery of Kibbutz Gan Schmuel, Israel.

estimates of egg numbers can be obtained. The eggs are then placed in upwelling hatching containers to incubate for 18 to 28 hr for grass carp and 43 to 55 hr for common carp, depending on water temperature[219] (Figure 4).

When the eggs hatch the larvae swim to the surface and are carried by water currents into larval rearing tanks. Three to five days after hatching the yolk is totally absorbed, the fry surface to gulp air and inflate their swimbladders, and feeding begins. Food such as decapsulated brine shrimp nauplii, rotifers, and yeast are fed on a continuous basis.[201,218] After about 10 to 14 days in the hatchery the fry are removed from the tanks and placed in nursery ponds.

C. Genetics

Since the common carp has been domesticated longer than any other fish it has been possible for culturists to undertake genetic studies and develop several different strains and two major races.[201,227-238] The European race of common carp consists of four main strains characterized by different scale patterns: (1) the "scaly" carp is totally covered with scales, (2) the "mirror" carp has only a few scales under the dorsal fin and ventrally, (3) the "line" carp has scales only along the lateral line, and (4) the "leather" carp has no scales. The leather and line carp varieties carry recessive lethal genes and are characterized by slow growth and low fecundity. Many highly selected and inbred lines of mirror and scaly carp have been developed in Europe and Israel.

The process of domestication of the common carp in Europe has led to the development of highly inbred lines with concomitant increases in deformities and reduced growth rates. The most common practice in Israel to counter these undesirable traits is to cross the line known as the "Dor-70" with a Yugoslavian line to restore hybrid vigor.[201,236-238]

The Asian race known as the "big belly" carp was domesticated under totally different conditions than the European carp. The former is a minor component of inten-

FIGURE 4.   Carp eggs in incubators at the fish hatchery of Kibbutz Gan Schmuel, Israel.

sively manured polyculture systems as compared with the European race which is usually reared in monoculture. The Asian or Chinese variety of common carp has not been
highly selected or inbred and retains the wild characteristic of high fecundity, but
grows relatively slowly.[234,238] The high fecundity of that fish is demonstrated by the
large gonads of both sexes which give rise to the common name for the race.

The big belly carp is more difficult to seine than other carps because of its ability to
escape under nets. Crosses between the European and Asian carp have produced hybrids which grow more rapidly than the Asian race, have higher disease resistance, are
easier to seine, yet compete well in intensive polyculture situations.[232,236,238]

Research to produce gynogenetic monosex female and sterile triploid common carp
has led to populations with 5 to 10% faster growth compared with normal populations.[239,240] Gynogenetic grass carp production has also been studied in attempts to
produce monosex fish which would, when stocked, pose a reduced environmental
threat since spawning could not occur.[241-243] In addition, female grass carp have been
crossed with big head carp males to produce sterile triploid fish, though possibly fertile
diploids also occur as a result of such crosses.[244-246]

VI. FEEDING AND NUTRITION

A. Natural and Live Foods

Common carp fingerlings and adults are primarily benthic omnivores which feed on
chironomid larvae and adults, oligochaetes, and gastropods.[247-251] In cases where food
is limited carp have been found to eat tilapia fry, helping to reduce unwanted reproductive success of tilapia in polyculture ponds.[248]

Carp are fed silkworm pupae in Japan where those organisms are available in large
quantities as a by-product of the silk industry.[26,204] Carp farming in Japan started in
the silk-producing areas where pupae were readily available and inexpensive.

Larval and fry carp feed on zooplankton in nature.[203,252,253] Larval carp require small food organisms such as rotifers for about their first week on feed, after which they consume copepods and cladocera. Fertilized fry ponds can be managed to produce primarily rotifers initially by treatment with 1.0 ppm of organophosphorus esters such as Flibol E, Ditrifon, or Dipterex, which selectively kill arthropods at the recommended concentration.[252-253] These compounds degrade within a week allowing the copepods and cladocerans to become reestablished at the time the switch in primary food organisms is made by the fish. Under intensive hatchery conditions, larval and fry carp are often fed brine shrimp nauplii.[218,254]

Adult grass carp are herbivorous. They feed on submerged vascular plants and terrestrial plants that come in contact with the water.[201,255] In pond polyculture, terrestrial plants are sometimes cut and thrown into the water as supplemental feed for grass carp.

Young grass carp, like common carp, initially feed on zooplankton.[256-259] They begin feeding on rotifers and then alter their food habits to take crustacean and chironomid larvae. By 50 mm total length they become almost exclusively herbivorous.

B. Fertilization and Manuring

Natural productivity in fish culture ponds will generally support polyculture production levels of 20 to 40 kg/ha. The use of moderate amounts of organic or inorganic fertilizers can increase production to 200 to 400 kg/ha. This increase is accomplished by a concomitant increase in primary and detrital production which in turn increase the amount of natural food available to the carp. Fertilizers may be the only nutrient source applied to ponds in Europe.[26] They are also used as relatively inexpensive food-producing materials in conjunction with intensive monoculture or polyculture systems when combined with the use of pelleted feeds.[201] Heavy manuring has also been employed as the only source of nutrients in intensive carp polyculture.[205,206] There is ample literature available on the experimental use of inorganic and organic fertilizers in carp culture.[201,203,204,215,260-263]

C. Feeds and Feeding

The composition of pelleted carp diets has been widely reported.[21,26,201,204,264-273] Research has been conducted on the use of steroid hormones as growth-promoting feed additives,[274-277] and some work has been done on the development of prepared feeds for young carp.[278-281]

Optimum feeding rates for carp vary depending on fish size and age, water temperature, relative amount of natural food available, and the quality of the feed being offered. Feeding rates for production ponds in Israel are generally 4 to 5% of body weight daily.[201] Carp are fed either once in the morning or continuously with demand or automatic feeders. Some attempts have been made to feed grass carp pelleted diets, but that practice is not common.[282,283]

D. Nutritional Requirements

The nutritional requirements of carp have been widely researched and are well documented.[284-322] Those requirements have been the subject of two recent review publications,[21,323] and the interested reader is directed to those sources for further information.

VII. DISEASES

One of the most common and important diseases of the common carp is infectious carp dropsy,[201] also known as bacterial hemorrhagic septicemia and infectious abdominal dropsy.[324] It is primarily caused by the bacterium *Aeromonas punctata* (synonymous with *A. liquefaciens*), though it is often found to occur in conjunction with *Pseudomonas fluorescens.* The most obvious sign of the disease is accumulation of clear fluid in the abdominal cavity which leads to distension of the abdomen. Other signs include ulcers in the liver, decreased erythrocyte numbers and consequent lower hemoglobins, increased leukocytes, and lesions in the intestinal tract.[324] The disease occurs after the carp have been stressed. It is most common in the early spring after a hard winter when the stored food reserves of the fish have been depleted. Carp with high levels of blood and serum proteins resist exposure to the disease.[324]

Infectious carp dropsy can be prevented through employment of the following prophylactic measures:

1.   Thorough drying of rearing ponds during the winter to remove carrier wild fish and parasites
2.   Fertilization of rearing ponds to establish a good natural food supply
3.   Stocking of only healthy fish into rearing ponds
4.   Adequate provision of feed in the fall to allow the build-up of large nutrient reserves (primarily fat) in the fish
5.   Injection of fish during the spring with antibiotics such as chloromycetin or streptomycin

Columnaris disease, caused by the bacterium *Flexobacter columnaris,* is a common fish pathogen which has recently been reported from carp in Europe for the first time.[325] Virulent strains produced external ulcers surrounded by red areas while other strains produced different signs. Successful treatment was accomplished with penicillin and ampicillin.

Research in Israel[230] has shown that the susceptibility of carp to certain diseases is related to recessive genes. Fish pox, epidermal epithelioma, and swim bladder inflammation have all been shown to occur only in fish which were homozygous for certain recessive traits and were members of highly inbred lines. The use of crossbreeding in Israel has greatly decreased the incidence of the diseases mentioned.

External parasites are another common problem.[201] The monogenetic trematodes *Dactylogyrus* and *Gyrodactylus* are problems in fry but can be controlled with 0.2 ppm of Bromex in ponds. The parasitic copepods known as fish lice (*Argulus* sp.) and anchor worms (*Lernaea* sp.), while mainly problems of esthetics in adult carp, can be fatal when present in high concentrations on young fish. Fish lice can be controlled with low levels of Lindane, Malathion, or Bromex in ponds. Adult anchor worms attached to carp are resistant to pesticide treatments but the free-swimming larvae can be killed with Bromex at levels of 0.12 to 0.15 ppm active ingredient in ponds. The chemical should be applied three times at 1-week intervals during summer. Anchor worms and fish lice can also be controlled with 0.2 to 0.3 ppm Dipterex applied to ponds at 2- to 3-week intervals.[204]

The protozoan *Ichthyophthirius multifilliis* can also be a problem in carp culture.[204,324] It can be eliminated by raising the water temperature above 25°C or by treating ponds with either 0.10 to 0.15 ppm malachite green or 20 ppm formalin. The fungus *Saprolegnia* sp. is commonly seen as a secondary infection during the winter when carp are weakened following injury or handling stress. The fungus can be controlled prophylactically in rearing ponds by the application of 25 ppm formalin.

VIII. BUFFALO CULTURE

There are three species of buffalo native to North America: bigmouth buffalo (*Ictiobus cyprinellus*), smallmouth buffalo (*I. bubalus*), and black buffalo (*I. niger*). Of these the bigmouth buffalo is the primary species considered for culture because it grows most rapidly, is best adapted to pond and lake environments, matures earliest, and has a higher fecundity than the other species.[326] It can be distinguished from the smallmouth and black buffalo by the location of its mouth, which is terminal, while the mouths of the other two are inferior. The bigmouth buffalo feeds primarily on zooplankton.

Buffalo are raised primarily in Arkansas and nearby states either in monoculture in fertilized rice field ponds undergoing a fish cycle of crop rotation or in polyculture with channel catfish.[206] The recent Payment In Kind (PIK) program initiated by the federal government has produced some shift in acreage from grain to bigmouth buffalo.

The bigmouth buffalo is spawned and the fry reared in a manner similar to that described above for common carp.[326-328] The fish spawn in the early spring when water temperatures rise from 16 to 21°C. They spawn during flooding periods and lay their adhesive eggs on vegetation.

Brood fish of 1.5 to 4.0 kg are selected in the late fall or early winter from the most rapidly growing stock. They are treated prophylactically to remove external parasites, and are placed in overwintering ponds at high density (500 to 1000 kg/ha) to inhibit premature spawning in the spring. Spawning ponds are prepared by draining to kill diseases and predatory wild fish. Grass is allowed to grow in the ponds to provide spawning substrate. Spawning ponds are filled immediately before the brood fish are stocked. Spawning generally occurs within a few days. If the fry are to be transferred to a nursery pond, 18 to 22 pairs of brood fish are stocked per hectare of the spawning pond. If the spawning pond is also to be used as a nursery pond, only three to five pairs of brood fish are stocked per hectare. Males can be differentiated from females by the rough sandpaper feel of their scales during breeding season.

If spawning has not occurred within 2 weeks of stocking, the ponds can be drained and refilled. This will sometimes induce the fish to spawn. Brood fish can be stripped in the hatchery and the wet method of fertilization used with stirring for 1 hr in a corn starch solution to prevent adhesion until the eggs water harden. The fish can then be hatched in jars.[327] Within the range of 16 to 21°C, eggs will hatch in 5 to 10 days.[326-328]

Spawning ponds are fertilized at the time of spawning to provide food for the larvae following yolk sac absorption. Ponds can be periodically treated with oil to kill predatory insects. At 10 to 15 days, supplemental feeding with finely ground fish feed can be implemented, particularly in instances where the fry are observed to school along the edge of the pond in search of food. At 3- to 5-cm long the fry can be seined and moved to secondary nursery ponds. Such ponds are commonly stocked at 4500 to 22,000 fish/ha and are fertilized at frequent intervals to provide natural zooplankton. Fingerlings for subsequent stocking can be produced by the end of the first 8-month growing season.

Buffalo are usually grown out in extensive ponds ranging in size from 5 to 100 ha. The final market size of fish desired controls stocking density; the larger the fish needed the lower the stocking density. The range is generally from 50 to 150 fish/ha.[326,328] Monoculture ponds are either fertilized only, or are fertilized and provided with supplemental sinking feed. In polyculture with channel catfish, buffalo are expected to feed off natural foods and excess catfish ration. Polyculture research has been conducted with combinations of bigmouth buffalo, channel catfish, crayfish, and paddlefish[329] as well as with hybrid bigmouth x smallmouth buffalo, common carp, tilapia, and channel catfish.[247]

Buffalo are marketed at 1 to 3 kg. The larger fish sometimes bring a higher price per kilogram than the smaller ones.

Chapter 4

TILAPIA

Robert R. Stickney

TABLE OF CONTENTS

I. INTRODUCTION

Tilapia are members of the family Cichlidae. In body configuration they are, in general, similar to sunfishes (Figure 1). About ten species have been successfully cultured,[330] and as a group tilapia represent one of the most widely produced foodfishes in the world at the present time (surpassed only, perhaps, by carp).

While the family Cichlidae is widely found throughout tropical and subtropical latitudes, the original distribution of tilapia was from south central Africa northward into Syria, with populations also occurring on the island of Madagascar.[331,332] Tilapia first appeared outside their native range when they were introduced to Java by persons unknown in 1939.[333] Further dispersion around the islands of the Pacific Basin was carried out by the Japanese during World War II.[334] Because of successes in the culture of tilapia following the war, and with the development of the theory that the ease of culture and high productivity of the fish could contribute significant amounts of animal protein in developing countries, the fish were further distributed throughout the Indo-Pacific and into both North and South America.[335]

More complete information on distribution and habitat preference for various species of tilapia is available in a paper by Philippart and Ruwet.[336] Tilapia and their culture have been widely reviewed in recent years.[337-340] In addition, a bibliography on tilapia was published in 1982.[341]

Recent changes in the taxonomy of tilapia have been suggested,[3] but as indicated in Chapter 1, the generic name *Tilapia* is adhered to in this text in accordance with the American Fisheries Society.[2] The reader should be aware that much of the literature utilizes the generic names *Sarotherodon* and *Oreochromis* in place of *Tilapia* for certain animals currently under culture.

Perhaps the most widely cultured species of tilapia are *T. aurea* (Figure 2), *T. mossambica,* and *T. nilotica.* Of the three, *T. mossambica* appears to be the least desirable, though since it was the first to be distributed outside of Africa and the Middle Eest, that species continues to be one of the most widely cultured. Both *T. aurea* and *T. nilotica* grow more rapidly than *T. mossambica* and reproduce at a later age and, consequently, larger size. The native distribution of *T. aurea* overlaps that of *T. nilotica* in Senegal and Chad,[336] so some physical or biological requirement leading to ecological isolation in the wild is indicated. Both species were historically present in the Nile River and *T. aurea* is native to portions of Israel.[342] Because of hybridization, pure stocks of many species are difficult, if not impossible to find, even in their native regions. Electrophoresis has shown that fish thought to be *T. nilotica* in Israel, for example, are actually some type of hybrid.[343]

Hybrids formed from various species are also currently being reared in some regions. One is the so-called red or golden tilapia which is touted as being more easily marketable than the pure strains which often have dark coloration and a black peritoneum. The red tilapia appears to have been initially produced in Taiwan as a cross between *T. mossambica* and one or two other species. The hybrid is fertile, but breeding between hybrid pairs leads to a wide variety of colors, including those of the fish which produced the hybrid.

Because of the intolerance to cold (discussed below), tilapia culture will be limited in the continental U.S. to extreme southern regions and to those isolated situations in which warm water is available throughout the year (geothermal regions and in the cooling water of power plant lakes). Through proper overwintering of broodstock and/or fingerlings, summer production of the fish is also possible throughout much of the south and southwest. Attention on these fish in this volume is warranted, in any event, because tilapia have such a great impact on warmwater fish culture around the world.

FIGURE 1.   Fingerling and marketable *Tilapia aurea.*

II. WATER QUALITY REQUIREMENTS

Tilapia as a group are some of the most hardy fishes known. They are tolerant of both crowding and poor water quality, features which make them excellent candidates for culture when coupled with their excellent flavor, rapid growth, and the fact that they feed low on the food chain.

A. Temperature

While tilapia are able to tolerate relatively high temperatures, the species presently under culture are tropical and have only a limited ability to survive cold. When exposed to cold water, disease resistance is severely impaired and death may result in a matter of hours or days, depending upon the species and actual temperature (Table 1). Under conditions a few degrees above lethal, tilapia rapidly develop such things as fungal and bacterial infections which contribute to high mortality even if temperatures do not become reduced enough to cause direct mortality.

As a general rule, tilapia do not grow well at temperatures below about 16°C and cannot usually survive for more than a few days below about 10°C.[338] The normal range of temperature over which *T. aurea,* a representative species, is found is from about 13 to 32°C.[336]

Growth of such species as *T. mossambica* is three times more rapid at 30°C (approximately optimum temperature for growth) than at 22°C.[334] The optimum temperature

FIGURE 2.   Rapid growth of tilapia can be obtained when organic wastes are added to ponds. In this case, 200 laying hens were maintained over a 0.05-ha pond stocked with *Tilapia aurea* at 1/m².

Table 1
MINIMUM TEMPERATURE
TOLERANCES FOR SOME
SPECIES OF TILAPIA

Species	Lethal low temperature (°C)
Tilapia aurea	8—9
Tilapia mossambica	8—10
Tilapia rendalli	11
Tilapia sparrmanii	7
Tilapia vulcani	11—13

Developed from information reviewed by Chervinski, J., in *The Biology and Culture of Tilapia*, Pullin, R. S. V. and Lowe-McConnell, R. H., Eds., International Center for Living Aquatic Resources Management, Manila, 1982, 119.

for feeding of *T. zillii* is from 28.8 to 31.4°C,[345] a range which may also be representative of other species.

Tilapia can tolerate temperatures of over 40°C in some instances, though there are also reports of mortality occurring at about 38°C.[338,346-348] Acclimation time and temperature may have a bearing on subsequent upper lethal temperature.[344]

T. aurea seems to be one of the most cold tolerant of the cultured species of tilapia, with a generally accepted lower lethal temperature of between 8 and 9°C[349,350] (Table 1), though one report indicated that the species could tolerate as little as 6°C.[351] *T. mossambica* is probably typical in terms of its response to cold. That species ceases feeding at about 16°C, and death begins to occur at between 11 and 14°C. Total mortality occurs in the range of 8 to 10°C.[352]

In tropical environments where ambient water temperature may vary only slightly over the course of a year, few problems with temperature tolerance are encountered by tilapia culturists. However, in subtropical and temperate climates, winter conditions can be expected to lower water temperatures to lethal levels for tilapia. To prevent winterkill, some source of warm water must be provided. Geothermal water sources and the thermal effluent from power plants have been utilized to some extent for tilapia rearing, but their use could be much more fully developed. Given a temperature gradient, such as that which occurs in the discharge effluent from a power plant, tilapia will readily find the warmest water and remain within it until general warming of the water body occurs in the spring. Thus, power plant lakes can support self-sustaining populations of tilapia even in regions which have relatively cold winters, provided the power plant does not shut down during cold weather.

When abundant supplies of geothermal water are not available and the culturist does not have access to the heated effluent from a power plant, other options must be examined. The problem is simply one of providing suitable conditions to carry fish through the winter. For reasons of economics, it makes the most sense to overwinter brood stock, small fingerlings or both, though in Israel it is standard practice to overwinter intermediate size fingerlings for final growout the following spring.[201]

Tilapia can be held in covered ponds which receive a constant supply of well water of sufficient temperature to keep them alive and healthy or they may be held in ponds, tanks, or raceways constructed under greenhouses.[353,354] Well water of suitable temperature should be constantly flowed into the culture chambers if no other source of supplemental heat is available, otherwise diurnal temperature fluctuations may lead to fish mortality. A properly designed greenhouse system can provide for maintenance and even spawning of broodstock during the winter, allowing the culturist to have a supply of young fingerlings prepared for stocking as soon as ambient pond water temperatures allow.[355]

B. Salinity

While tilapia do not appear to occur in saltwater within their normal range, tolerance to marine waters is, at least in some cases, remarkable. Reports of large populations of tilapia becoming developed in coastal waters following their introduction as exotic species have been appearing in recent years. Perhaps the most striking example is in Tampa Bay, Fla., where *T. melanopleura* is now the dominant fish species present, supporting a large commercial fishery. The impact of such population explosions on native fishes is of concern to marine ecologists, though such problems appear to be rare at the present time. Studies to determine the conditions necessary to allow tilapia to outcompete native fishes in coastal regions are certainly warranted if the problem is to be avoided in the future.

Growth of such species as *T. aurea* in natural or artificial seawater (in ponds and in the laboratory, respectively) is similar to that obtained in freshwater environments.[356,357] The ability of *T. zillii* to acclimate to and grow in seawater is even better developed than that of *T. aurea*.[358] Hybrid tilapia (a cross between a *T. aurea* male and a *T. nilotica* female) have been shown to perform well in brackish water ponds in Israel.[359,360]

According to Chervinski,[338] neither *T. zillii* nor *T. aurea* is able to reproduce in seawater. While no nest building has been shown to occur in seawater ponds containing *T. aurea*,[356] nesting behavior by that species has been observed at various salinities in aquaria.[357] In a pond study, the gonadosomatic index (relationship between gonad weight and fish body weight) of *T. aurea* dropped in fish reared in seawater,[358] but in aquaria, developing ovaries were found at salinities from 0 to 35 ppt (the experiment was conducted at 5 ppt salinity intervals) with the exception of 25 ppt.[357] Further study is certainly warranted.

At least one species, *T. mossambica,* will reproduce over a salinity range from fresh to full strength seawater, with one report indicating that reproduction occurred at 49 ppt salinity.[361] The same species has been shown to survive salinities as high as 69 ppt,[362] but there have been no reports of *T. mossambica* becoming widely established in marine environments.

C. Dissolved Oxygen

The ability of various species of tilapia to survive low levels of dissolved oxygen (DO) is widely recognized. The lowest DO level at which survival has been reported is 0.1 mg/ℓ. That value has been recorded for both *T. mossambica*[363] and *T. nilotica*.[364] A slightly higher value, 0.3 mg/ℓ, was reported for hybrid tilapia (*T. hornorum* male x *T. nilotica* female).[365]

In situations where high levels of primary productivity are present in association with tilapia rearing activities, occurrences of extremely low levels of DO are not unusual. DO may frequently approach or reach 0.0 mg/ℓ during the predawn hours. Those conditions may occur day after day over periods of many days or even weeks under the proper conditions, and while they would prove fatal to most fish species, tilapia are frequently not only able to survive, but will also continue to grow at a relatively normal rate.

Tilapia seem to have a well-developed ability to obtain oxygen from the saturated water which occurs at the air-water interface. During periods when the water column is depleted in oxygen, the fish will gulp at the surface, passing the oxygenated water over their gills. Mortality in *T. aurea* has been reported as a result of low DO in one instance where a pond was covered with a heavy growth of duckweed (*Lemna* sp.). The presence of the duckweed precluded the opportunity for the tilapia to obtain oxygen from the surface microlayer.[366]

D. Ammonia and pH

Few studies on the tolerance of tilapia for environmental ammonia have been conducted, but it is generally felt that these fishes are more tolerant than many others. Redner and Stickney[367] found that *T. aurea* which had not been chronically exposed to elevated ammonia levels had an LC_{50} (concentration which was lethal to 50% of the exposed fish) of 2.4 mg/ℓ unionized ammonia. When the same species was expoosed to sublethal unionized ammonia concentrations (0.4 to 0.5 mg/ℓ) for 35 days, the LC_{50} value was raised to 3.4 mg/ℓ. That study showed that acclimation conditions affect the response of the fish to the environmental contaminant, a concept which may be shown in the future to have broader application among other fish families.

The pH of freshwater fish ponds is generally within the range of 6.5 to 8.5 and is controlled by the carbonate-bicarbonate buffer system.[1] In poorly buffered waters, relatively broad fluctuations in pH can occur when high levels of primary productivity are present since carbon dioxide levels will vary considerably as a function of photosynthetic as opposed to respiratory rates in ponds. During the daylight hours, photosynthetic activity removes carbon dioxide from the water causing the pH to rise. Conversely, at night when respiration occurs in the absence of photosynthesis, carbon dioxide is released to the water to form carbonic acid which causes a reduction in pH.

Table 2
PRODUCTION LEVELS OF TILAPIA, COMMON CARP, AND CHANNEL CATFISH UNDER CAGE CULTURE CONDITIONS

Species	Estimate annual production (kg/m³)	Feed type (% protein)	Food conversion ratio
Tilapia aurea	200 to >300	40	1.0—1.8
Tilapia nilotica	200 to >300	25—30	1.9—2.2
Cyprinus carpio	Over 240	30—35	1.6—2.3
Ictalurus punctatus	Over 240	36	1.4—2.5

Developed from information reviewed by Coche, A. G., in *The Biology and Culture of Tilapia*, Pullin, R. S. V. and Lowe-McConnell, R. H., Eds., International Center for Living Aquatic Resources Management, Manila, 1982, 205.

Tilapia commonly occur in environments which have high levels of primary productivity; thus, the fish may be exposed to relatively high diurnal pH fluctuations. The tolerance of tilapia, in general, has been reported to be over a pH range of 5 to 11.[338]

III. PRODUCTION

Tilapia can be grown under a variety of circumstances, the most common of which is in ponds. Because of their tolerances to crowding and degraded water quality, they are most commonly reared in monoculture, though polyculture has also been practiced, particularly when tilapia is a species of secondary importance.

Yields of tilapia in ponds have been shown to positively correlate with phytoplankton numbers, Secchi disk transparency, gross primary productivity, and the level of chlorophyll a in the water.[368] The best correlation among those variables was for chlorophyll a (r^2 = 0.89). Since tilapia feed low on the food chain (most species consume plants, zooplankton, or both), the positive correlations between fish production and the indicated variables seems reasonable.

Stocking rates for tilapia are highly variable, depending upon the type of culture technique employed. In regions where subsistence culture is practiced and little or no supplemental feed is provided, stocking densities as low as a few hundred fish per hectare may be utilized. More intensive pond culture, such as that practiced in developed countries, typically employs stocking densities of 1/m² or higher. If some water exchange is provided within a pond, several fish/m² can be stocked and easily reared to market size. Representative stocking densities for tilapia in ponds range from 3000 to 5000 fish/ha in polyculture through 10,000/ha as a fairly average monoculture density, up to as many as 60,000/ha.[201,365,369]

The practice of culturing tilapia in ponds seems to have been developed most highly in Israel where several species have been under culture for many years. Nursery ponds are often utilized as an intermediate stage between spawning and production ponds under the Israeli system,[201] though that technique is not commonly applied in other countries.

Intensive tank and raceway culture can lead to production levels higher than those achieved in ponds, though cage production has been similar to that of high intensity pond culture. In a study which compared the growth of hybrid tilapia (*T. mossambica* x *T. hornorum*) in cages within ponds and in a laboratory, it was estimated that production could reach as high as 50,000 kg/ha/year.[370] Coche[340] compared the production of tilapia in cages with that of other species and showed that there were no great differences with respect to species (Table 2).

Growth, condition factor, and food conversion ratio (measured as weight of feed offered/weight gain of fish) have been shown in the laboratory to be better for *T. aurea* than for *T. mossambica*. Various densities of both species were reared in flow-through tank systems and it was found that growth of *T. aurea* continued good when fish at the highest stocking density reached weights equivalent to 66.2 g/ℓ.[371] The *T. mossambica* began to die when their biomass reached about 20 g/ℓ, with death being attributable to the development of an autoimmune response.[372]

Tilapia of various species have been stocked in polyculture with common carp (*Cyprinus carpio*), mullet (*Mugil cephalus*), channel catfish (*Ictalurus punctatus*), and freshwater shrimp (*Macrobrachium rosenbergii*) as a means of increasing total overall pond production.[201,248,373-376] If *T. aurea* are stocked with common carp at a density of no more than 5000/ha, they will not adversely affect, and might even stimulate common carp growth.[248] The same study demonstrated that tilapia growth was not affected by the presence of either carp or mullet at densities of 2500 to 3000/ha. Part of the success of polyculture relates, at least in some cases, to the fact that other fish species may prey upon tilapia fry as they are produced, thus reducing the competition for food by the products of unwanted reproduction. This subject is considered in more detail below.

IV. FERTILIZATION

Since the cultured species of tilapia feed low on the food chain and most feed to as large extent by filtering food organisms from the water through their gill rakers, the provision of natural foods through fertilization programs is a logical management tool. Inorganic fertilizers, organic wastes, and a combination of the two have been widely utilized to provide natural food for pond-reared tilapia. Organic fertilization has long been utilized as a primary mechanism to provide food not only for tilapia but also to various species of carp and other fishes in the Far East. The technique is also employed in commercial aquaculture in Israel[201] and other countries in the Middle East, Europe, and elsewhere.

In the U.S., pond fertilization has depended primarily on inorganic fertilizers for the production of plankton blooms. Potential difficulties with the utilization of such organic fertilizers as swine and poultry manure because of potential public health problems have influenced the amount of attention given to the subject by researchers. Some research with both inorganic and organic fertilization with respect to tilapia production has been conducted in the U.S. in recent years, though at present, fertilization in commercial operations seems to be restricted to inorganic compounds.

Boyd[377] examined the effect of 15 biweekly applications at the rate of 22.5 kg/ha of three inorganic fertilizers on the production of *T. aurea* in Alabama. The fertilizers were 0-20-5, 5-20-5, and 20-20-5 (N-P-K) compounds. Increased fish production occurred as the level of nitrogen in the fertilizer was increased because of the influence of nitrogen on phytoplankton production. In general, the types of fertilizer utilized in tilapia ponds are similar to those used in channel catfish production, though the frequency and application rates may be increased in tilapia culture over those which predominate in the catfish industry.

With respect to organic fertilizers, the most commonly utilized materials are poultry and swine wastes, though cattle waste has also been evaluated.[378] The recycling of organic wastes through fish has been reviewed by Edwards.[379] While most of the emphasis has been placed on establishing fish culture systems to which organic fertilizers are added in the interest of producing fish for human food, the use of tilapia in removing algae from sewage treatment lagoons is also a possibility. With the latter type of system, sale of the fish for human consumption in the U.S. may be prohibited, but the use of fish could actually be less expensive than other forms of secondary treatment in certain instances. In such cases, the fish could be discarded rather than being sold. The removal of microalgae from sewage systems has received some attention by researchers.[380-382]

Various studies have been conducted during recent years in Texas with tilapia reared in ponds which received the waste from either poultry (Figure 2) or swine.[366,383-386] Conclusions drawn from those studies were that the waste from the equivalent of 50 growing-finishing hogs per hectare or between 70 and 140 kg/ha/day of poultry manure will provide high levels of primary and secondary productivity and lead to excellent growth of tilapia. A similar figure for manuring relative to swine waste was derived by Aguilar.[387] With respect to poultry manure, best results in tilapia fingerling production in the Philippines is achieved when fertilization rates within the above range from poultry manure are employed.[388]

Mixtures of manure and sorghum were evaluated in Israel,[389] where it was determined that the natural food supplies were better in ponds which received the grain than in those which received only manure. It was felt that the *T. aurea* in the ponds were feeding on benthic animals and detritus to a large extent and that the production of those types of food was improved through the addition of the sorghum.

V. REPRODUCTION

Most of the commercially cultured species of tilapia are mouthbrooders. The strategy involves nest building activity by the male followed by a brief period of courtship which results in spawning. Eggs are laid in the nest, fertilized, and picked up for incubation in the mouth of one of the adults (in most species the female incubates the eggs). Once the fry have hatched, they remain in the mouth of the female through yolk sac absorption. Swim bladder inflation will also generally occur during the posthatch period when the fry are under the protection of the female.[390] For a few days after the fry are released, the female will remain in close proximity and the young fish will seek refuge in her mouth when danger threatens.

While the fecundity of an individual tilapia female is relatively low (for example from 69 to 302 eggs per female in *T. galilaea*[391]), the onset of spawning at an early age coupled with a short refractory period lead to sexual maturity at a size as small as 10 cm,[333] and the production of as many as 6 to 11 broods wihin a year under proper environmental conditions.[334] Onset of spawning may occur at an age of 3 months in *T. mossambica*. Other species, such as *T. aurea* and *T. nilotica,* generally begin spawning at about 6 months. *T. nilotica* females mature at 11.4 cm and males at 14.3 cm on the average.[392] Once the females of any mouthbrooding species being to mature, energy is diverted from growth to egg production and a distinct divergence with respect to size begins to appear in a population of fish of the same age.

Another strategy involves substrate spawning, where the eggs remain within the nest during incubation and yolk sac absorption. Fecundity in such substrate spawning species as *T. zillii* may be over 1000 eggs on the average, but the fish are initially smaller than those of mouthbrooding species and growth is not as rapid, so mouthbrooders are more popular among culturists.

In addition to growth reduction during gonadal development and following maturity in mouthbrooding tilapia females, a secondary problem of overpopulation and consequent stunting seems to be common. Reproduction leads to increased competition for food within a pond and can lead to degradation of water quality. Strategies have been developed, as discussed below, to reduce or eliminate spawning in growout situations.

FIGURE 3.   Spawning happas are commonly used in the Philippines for reproduction and fry rearing of tilapia.

A. Captive Breeding

Spawning will occur in ponds, tanks, cages, and other types of culture environments. Photoperiod does not seem to be important, but temperature is critical. Tilapia will not spawn if the temperature falls below from 20 to 23°C;[393,394] thus, in temperate climates spawning begins in the spring and may continue until early fall. In the tropics, year round spawning may occur.

The simplest technique for spawning tilapia involves stocking broodfish in an open pond and later collecting fry or fingerlings. Once the fry leave the mouth of the brooding female, they swim in schools for several days, often appearing at the surface near the edge of a pond where the water is warmest. Such schools can be easily netted. However, as the spawning season progresses, individuals and schools of young fish which escape capture will reach sufficient size that they can cannibalize newly released fry. Cannibalism may be the reason for the disappearance of fry in brood ponds after several weeks of spawning. The phenomenon has been documented in aquaria and tanks.[393,395-397]

Pond spawning has been widely used in the production of hybrid tilapia. Such crosses as *T. mossambica* x *T. nilotica* can be readily produced[398] and may grow more rapidly than either of the parents. The golden tilapia (a hybrid between *T. mossambica* and *T. hornorum*) has recently become a popular hybrid in many parts of the world. The obvious problem with hybridization is that the species used in the cross must be maintained separately until paired for spawning, and a high degree of precision must occur in conjunction with sex determination to ensure that both sexes of one of the species are not stocked in the same breeding pond.

The happa spawning method is widely employed in the Philippines. A happa is a small net-covered cage of mesh size fine enough to prevent escape of the eggs during spawning (Figure 3). Several happas can be placed within a small pond and fry can be removed for stocking as they are produced. A modification on the happa technique involves the use of larger mesh netting or wire on one side of a happa-like cage. If the larger mesh will allow the fry to escape, the cage can be used to contain adults while reducing or eliminating cannibalism. The small cages can be moved from pond to pond for stocking fry as they are produced.

Tilapia can be spawned with relative ease in aquaria,[399,400] but that technique is not generally suitable for commercial culturists because of facility and personnel needs which would be associated with a high level of production. Aquarium studies have been important in outlining the spawning behavior[400] and environmental conditions[399] required by various species.

Stripping of females after development in aquaria or other types of confined culture systems has been accomplished,[399] followed by external incubation of the eggs. Eggs may also be obtained from the mouths of brooding females and incubated under artificial conditions. Shaker tables[399] and conical shaped incubation chambers with updraft water flow[401] have been successfully used as egg incubators. While there may be no particular difference in hatching success for eggs or fry collected from the mouth of females,[397] externally incubated eggs taken from the mouths of *T. aurea* females in one study had relatively low levels of survival, particularly when removed during the early stages of development.[355]

Ratios of males to females in spawning chambers may vary considerably, though it would seem to make sense to stock larger numbers of females than males. A male will quickly begin to court secondary females once he has spawned and the original female has picked up the eggs and abandoned the nest site. In tanks, a ratio of three males for each female has proven effective.[393,397]

Improvement of tilapia stocks through genetic control has not been widely addressed as yet, though there are areas where selective breeding and even genetic engineering might be expected to have significant impacts. Breeding for improved growth in *T. nilotica* was of only limited success because of low genetic variation.[402] Since many culturists are working with relatively small gene pools and there has been so much hybridization in the various stocks around the world, selective breeding may not be an effective means of improving performance. Gene splicing to improve tilapia stocks may be on the horizon, however.

One intriguing concept would be to introduce one or more genes which would improve the cold tolerance of the fish. While desirable from a culture standpoint, a cold tolerant tilapia could be objectionable in areas where the fish would compete with fish considered to be more desirable. For example, in the U.S., *T. aurea* have been found to outcompete largemouth bass (*Micropterus salmoides*) for spawning sites in a power plant lake where the tilapia were able to overwinter.[403] If tilapia were genetically engineered to become tolerant to temperate winters, similar competition could conceivably develop in a variety of other aquatic habitats, leading to significant reductions or even the elimination of largemouth bass populations.

B. Control of Reproduction

As reviewed by Guerrero,[404] reproduction in tilapia can be controlled by (1) ensuring that fertilized eggs do not survive, (2) stocking predators, or (3) rearing unisex populations. In the latter case it is, as indicated above, generally desirable to rear all-male populations because of their rapid growth as compared with females.[334,398,405-409] If reproduction is not controlled, stunted populations may occur as was recently demonstrated in *T. melanotheron* in West Africa.[410]

Early studies indicated that tilapia reared in cages could not reproduce. The lack of fry production was thought to be a function of altered reproductive behavior leading to the prevention of fertilization,[411] but instances of fry production have since been

Table 3

PREDATORY FISH SPECIES WHICH
HAVE BEEN USED TO ASSIST IN THE
CONTROL OF TILAPIA REPRODUCTION
AND THE REGION WHERE THOSE
SPECIES HAVE BEEN EMPLOYED

Predator	Location	Ref.
Cichlosoma manguense (Oscar)	El Salvador	414
Cichla ocellaris (Peacock bass)	Puerto Rico	415
Mudfish (species not given)	Thailand	335
Lates niloticus (Nile perch)	Africa	416—418
Dicentrarchus labrax)	Israel	419
Micropterus salmoides (largemouth bass)	U.S.	420

reported when cage mesh size was 0.3 cm or smaller.[412,413] The dependence on mesh size for successful reproduction is related to the size of the eggs being produced; if the mesh is sufficiently large, the eggs will fall through before the female can pick them up. The use of small mesh for egg retention is the key to production of tilapia fry in happas as discussed above.

Various species of fish have been stocked as predators on tilapia fry (Table 3). It is important that the growth of the tilapia originally stocked is rapid enough to outpace that of the predator; otherwise, the crop may be decimated. A properly selected predator can also represent a secondary crop. In most instances, the predator is selected from among locally available species; it is not generally considered wise to introduce an exotic predatory species.

Perhaps the most popular means of controlling reproduction in tilapia involves stocking all-male fish. This can be accomplished by hand sexing or through sex reversal. Hand sexing may be effective with fish as small as 40 g,[421] but fish from 50 to 70 g seem to be preferred.[339] The distinction between the sexes is more easily made if the genital papilla is stained with a water-soluble die such as Azorubin prior to inspection.[422] In any case, the technique is time consuming and not 100% sure. It is sufficiently reliable that it continues to be widely employed by commercial producers in Israel.

Sex reversal has been induced through hybridization of certain species (Table 4) and by feeding hormones to tilapia fry. Hybridization has been popular in Israel and the Far East, but as previously indicated, the accidental mixing of both sexes of the same species within a spawning pond can lead to problems. Further, the hybrids must be maintained separate from the parental species to preserve the integrity of the technique. Finally, the crosses are not always 100% effective (see review by Wohlfarth and Hulata[434]).

The most recent technique for sex-reversing tilapia involves feeding hormones to fry at swimup. While most workers have been interested in producing all-male populations by feeding androgens, some studies have focused on the production of all-females by feeding estrogens.[435,436] Successful production of all-male tilapia by feeding hormones has been reported for *T. mossambica*,[437,438] *T. nilotica*,[439-441] and *T. aurea*.[355,442,443] The technique described by Guerrero[442] is representative.

Guerrero found that when *T. aurea* fry were fed for about 3 weeks with a prepared diet containing 60 mg/kg of 17-alpha-ethynyltestosterone, only male fish were pro-

Table 4
RESULTS OF *TILAPIA* HYBRIDIZATION FOR
THE PRODUCTION OF PRODUCTION OF
UNISEX POPULATIONS

Species involved in cross	Ref.
Crosses leading to predominantly unisex populations	
T. nilotica x T. hornorum	423,424
T. mossambica x T. aurea	396,398,425,426
T. hornorum x T. aurea	417,427,428
T. nilotica fem. x T. aurea male	429
T. vulcani fem. x T. aurea male	429
Crosses leading to all-male offspring	
T. mossambica fem. x T. hornorum male	423,424
T. nilotica fem. x T. hornorum male	417
T. nilotica x T. macrochir	430
T. nilotica fem. x T. aurea male	417,431
T. nilotica x T. variabilis	417
T. spilurus niger x T. hornorum	417
Cross which leads to all-female offspring	
T. melanotheron fem. x T. mossambica male	380,381

duced. The feed is prepared by dissolving the appropriate amount of androgen in ethyl alcohol and mixing the alcohol with the feed at a ratio of about 2 parts alcohol:1 part feed. The mixture is then placed in a drying oven at about 60°C until the alcohol has evaporated. During the evaporation process the androgen is adsorbed to the surface of the feed. When large numbers of fish are being sex-reversed the success rate may fall slightly below 100% because there is a chance that any given fish will not obtain sufficient amounts of hormone. Also, the level of hormone and duration of feeding which works well on *T. aurea* may not apply to other species.

VI. NUTRITION

Even though tilapia are among the most widely reared fishes in the world, relatively little attention has been directed toward outlining their nutritional requirements. The predominance of culture systems which rely on frequent fertilization as a means of providing natural food has certainly taken the emphasis away from prepared diets and thus, the need to clearly define the nutritional requirements of these fishes. Recently, fish nutritionists have turned their attention toward tilapia, however, and some significant strides have been made. A general feeding table for tilapia has even been proposed (Table 5).

A. Natural Foods

One way to begin evaluating the nutritional requirements of a fish or group of fishes is to examine their food habits. The composition of the preferred foods should provide some indication of how a prepared diet should be formulated relative to its protein, carbohydrate, lipid, and energy levels.

The ability of such species as *T. aurea* to grow rapidly on blue-green algae[385] and other primary producers has led to the view in some circles that tilapia of commercial aquaculture importance are herbivorous as juveniles in nature, though most species appear in reality to be omnivores (Table 6). Even species such as *T. zillii*, which has been introduced throughout much of the southwestern U.S. as a weed control species in irrigation canals consumes some animal matter in nature (Table 6), though it could be contended that animals are incidentally ingested with macrophytic material.

Table 5
RECOMMENDED
PREPARED DIET FEEDING
RATES (AS PERCENTAGE
OF BODY WEIGHT) FOR
TILAPIA

Average fish weight (g)	Feeding rate (%)
5—10	6.7
10—20	5.3
20—50	4.6
50—70	3.3
70—100	2.8
100—150	2.2
150—200	1.7
200—300	1.5
300—400	1.3
400—500	1.2
500—600	1.1

Adapted from Marek, M., *Bamidgeh,*
27, 57, 1975.

Table 6
FOOD HABITS OF SELECTED SPECIES OF TILAPIA

Species	Primary foods ingested	Ref.
T. aurea	Zooplankton, detritus	445,446
T. nilotica	Phytoplankton	447
T. mossambica	Macrophytes, benthic algae, phytoplankton, periphyton, zooplankton, fish larvae, fish eggs, detritus	448—453
T. galilaea	Phytoplankton, primarily *Peridinium cinctum* in Israel	446,454
T. zillii	Macrophytes, benthos	455,456

B. Protein and Energy Requirements

Studies to evaluate protein requirements and protein utilization by tilapia have employed a variety of feedstuffs, some commonly utilized in commercial fish feeds for other species and others of an exotic nature. Studies on *T. aurea, T. mossambica*, and *T. zillii* have indicated that each species performs best when dietary protein is in the 35 to 40% range.[457-459] However, hybrid *T. hornorum* x *T. nilotica* males showed no significant differences in growth on a series of diets which ranged from 20 to 35% protein.[460] There was no standardization of dietary protein quality or energy level in the cited studies, so the absolute protein requirements have not been completely outlined for any species as yet. The protein:energy requirements of *T. aurea* and *T. mossambica* has been examined[458,461] and appear to be similar, though more work in that area is also required.

To date the most concentrated effort on protein and amino acid utilization has been focused on *T. mossambica.* That species was shown capable of using free amino acids.[462] The quantitative indispensible amino acid requirements for *T. mossambica* have also been determined (Table 7).

It is possible to substitute such plant proteins as soybean, cottonseed, and feather meal,[457,464-466] though growth is generally depressed if excessive levels of any of those plant proteins is used to replace fish meal. In general, 25% replacement of animal

Table 7
AMINO ACID
REQUIREMENTS OF *TILAPIA*
MOSSAMBICA[462,463]

Amino acid	Minimum requirement (percent of diet)
Arginine	0.93
Histidine	0.37
Isoleucine	0.75
Leucine	1.28
Lysine	1.62
Methionine	0.16
Phyenylalanine	0.79
Threonine	0.95
Tryptophan	0.16
Valine	0.75

protein with plant meals appears to be acceptable. However, when the alga *Cladophora glomerata* was used to replace fish meal in diets fed to *T. nilotica*, growth and protein utilization were depressed at replacement levels above 5%.[467]

In developing countries such high quality plant proteins as soybean, corn, peanut, and cottonseed meal are either not available or are given priority for human or animal use and are not available for fish feeds. In addition, fish meal and other animal proteins may be unavailable or extremely expensive. Thus, there have been many attempts to find suitable alternative protein sources for tilapia diets in the developing world. Some have shown promise, while others do not promote very good growth. Such exotic feed ingredients as algae, ipil-ipil meal (*Leucaena leucocephala*), mulberry leaf meal, coffee pulp, copra, groundnut, distillation solubles, pharmaceutical wastes, spent beer waste, chicken feed, mosquito larvae, zooplankton, and lettuce have been used with variable success.[345,465,467-473] Poultry and cattle manure have also been incorporated into tilapia diets,[448,449] though as previously discussed those items are most commonly utilized in pond fertilization.

C. Carbohydrates

Few studies on the carbohydrate requirements of tilapia have been conducted, though it has been demonstrated that there is considerable potential for carbohydrate to spare protein in the diet of *T. mossambica*.[476] Diets high in fiber cannot be expected to have as much value for tilapia as those containing less complex carbohydrates since it has been shown that *T. aurea* do not harbor the enzyme cellulase.[477]

D. Lipids

Only recently has interest developed in outlining the lipid requirements of tilapia. In 1980, it was demonstrated that *T. zillii*, unlike many other fishes, may have a dietary requirement for fatty acids in the linoleic acid family.[478] Since then, research with *T. aurea* has indicated that linolenic acid may not be required by that species and further, that there may be interference with *de novo* fatty acid synthesis when linolenic acid is present at high dietary levels.[479] Supporting the difference between tilapia and such other fishes as trout and catfish is the fact that *T. aurea* grow as well on diets supplemented with soybean oil as fish oil. Beef tallow, which supports excellent growth in catfish, does not appear to be a very good lipid source for tilapia.[480]

E. Vitamins

Diets containing vitamin supplements developed for other species seem to perform well on tilapia, though deficiency symptoms can be induced, at least in the case of vitamin C. Signs of vitamin C deficiency in *T. aurea* include skeletal deformities similar to those which have been reported in other fishes.[481] It has been suggested that complete diets for *T. aurea* should contain a minimum of 50 mg/kg of vitamin C,[481] while the minimum requirement for *T. mossambica* appears to be about 80 mg/100 g of diet.[482] The reason for the apparent large difference between those species has not been determined.

No dietary vitamin B_{12} requirement has been demonstrated for *T. nilotica* since that vitamin can be synthesized in the intestine.[483] Both growth and erythropoiesis were normal in fish fed 16 weeks on a diet deficient in the vitamin. No decrease in the liver storage level of the vitamin was found over an eightfold weight increase.

F. Body Color

Commercial tilapia producers have become interested in the red tilapia hybrid and report that consumer acceptance is increased as compared with fish having normal coloration. There is a wide range in the color of hybrid tilapia, even when fish from the same spawn are compared. Carotenoid pigments have been found to enhance color in various species of fish, and tilapia are not exceptions. When lutein, rhodoxanthin, and the alga *Spirulina* sp. were fed to red tilapia there was color enhancement as compared with a control.[484] As yet, color-enhancing chemicals are not in standard use in the industry, but their use may increase in the future as more information becomes available.

VII. DISEASE

As indicated above, tilapia rarely demonstrate signs of disease except after exposure to cold temperatures. That does not mean, however, that the fish are immune when reared within their optimum temperature range. Various diseases have been reported for tilapia under culture conditions, including the virus *Lymphocystis*, bacterial problems with *Aeromonas*, myxobacteria, and *Edwardsiella tarda*, and such parasites as trematodes, trichodinids, and *Ichthyophthirius*.[485] Terramycin resistance has been reported in *T. aurea* for *E. tarda*.[173]

T. mossambica have been shown to display a hypersensitivity response to some component in the mucus under high density culture.[372] The unidentified component induced cutaneous anaphylactic reactions not only in *T. mossambica* but also in three other species of tilapia (*T. aurea*, *T. nilotica* and *T. zillii*). No reaction was elicited from channel catfish, however, so the component is thought to have utility in population control only in communities of closely related species.

Chapter 5

CENTRARCHIDS*

Bill A. Simco, J. Holt Williamson, Gary J. Carmichael, and J. R. Tomasso

TABLE OF CONTENTS

*   Use of tradenames in this chapter does not imply endorsement by the U.S. Fish and Wildlife Service nor the individual authors of the chapter.

I. INTRODUCTION

Initial demands for cultured black bass and sunfish (family Centrarchidae) were for stocking into natural waters as a means of establishing or supplementing existing sport fisheries. Federal hatcheries were established at the beginning of the present century to meet public demands for these sport fishes, and initiation of the Federal Farm Pond Program in the late 1930s added greater emphasis to sport fish culture.

Under the Farm Pond Program, bass were usually stocked at sizes of 2.5 to 5.0 cm. In that size range, stocking is often successful in new or recently renovated reservoirs where predation and competition from existing fishes are absent or greatly limited. In some cases, small fingerlings were stocked regardless of conditions in receiving waters and survival was highly variable. Nevertheless, the Farm Pond Program had a positive effect on the development of improved management methods.

Swingle and Smith at Auburn University pioneered research related to farm pond dynamics. Their work clarified predator-prey relationships for stocking combinations of largemouth bass and bluegills, species they recommended for southeastern farm ponds.[486-489] Broad implementation of their stocking recommendations contributed to increased demand for sunfish, leading to the production of millions of centrarchid fishes annually by federal, state, and private hatcheries.

Improved management programs stressed an integrated approach to restore pond balance and promote rapid growth of both predator and prey species with the goal of improving angling success.[490] Although largemouth bass has dominated production statistics, smallmouth bass, bluegill, and hybrid sunfish production have also been key elements of many stocking programs. Since basic culture techniques are similar among centrarchid basses, emphasis has been placed on largemouth bass production.

II. LARGEMOUTH BASS

Hatchery production of 3.8- to 5.0-cm largemouth bass (*Micropterus salmoides*) fingerlings has primarily been for the establishment of sport fisheries. Larger size fish have been used for corrective restocking, research studies, "put and take" fisheries, and have been maintained as brood stock. Production and stocking of advanced fingerlings [>100 mm total length (TL)] is an effective management technique to control established populations of forage fish, including sunfish. Lawrence[491] determined the size of bass required to prey upon bluegills of various sizes.

In the 1960s, Snow and associates at the U.S. Fish and Wildlife Service laboratory near Marion, Ala., developed the "Marion Program" which established the principles currently used in the culture of advanced fingerlings. Refinement of spawning and rearing techniques now permits intensive production of largemouth bass. Continued success depends on demand, as well as the originality and determination of managers.

A. Brood Stock

Broodfish have been historically collected from the wild by electrofishing or other means, but are now typically domesticated stocks that have adapted well to hatchery conditions. Initial selection of largemouth bass stocks may be strain dependent, as recent evidence indicates stocks differ substantially for important characteristics.[492] Maintenance of broodfish in good physical condition is important; healthy fish are resistant to environmental stresses and have high survival rates during winter. Some hatchery managers use a saline/antibacterial solution at times when broodfish are handled or hauled.

Sexual maturity depends on size rather than age.[493] In the southern U.S., a bass may mature at 8 months (22- to 24-cm long and 0.18 kg).[489] The time required to reach spawning size increases with increasing latitude except under unique conditions (e.g., fish in water receiving thermal effluent from power plants or geothermal spring water may mature earlier than fish at the same latitude which are maintained under prevailing ambient temperatures).

Random sampling should provide brood stock of both sexes since sex ratios tend to be equal in natural bass populations. However, male fish apparently do not live as long as females and sex ratios may be altered in older populations.[494] Selecting broodfish on external characteristics should be done with caution. For example, if size differences are correlated with sex, selecting broodfish based on size alone may lead to a brood population with a skewed sex ratio, thus seriously reducing effective population size.

Marking individual broodfish or fin clipping particular year classes or stocks and keeping accurate fecundity records are positive steps toward establishing proper breeding programs. Evaluation of blood and fin tissues by electrophoretic techniques now permits identification of Florida and northern strains of fish.[495] Such techniques should be valuable tools in bass selection programs.

Older broodfish should be replaced as necessary since spawning is less certain in them and they are larger and more difficult to handle than young adults.[496] Sexing of broodfish is easiest in the early spring prior to spawning. Females can often be distinguished by the presence of distended abdomens. Bass greater than 35 cm TL can usually be sexed by examining the scaleless area adjacent to the urogenital opening. That area is nearly circular in males but is elliptical or pear-shaped in females.[497] Ripe females are selected on the basis of an obviously distended and soft abdomen and a red, swollen, and protruding vent. The most certain way to sex fish is by insertion of a capillary tube as a catheter into the vent to detect the presence of eggs or milt.[494,498] Ripe males will emit sperm when stripped.

Proper care of broodfish should continue when spawning is complete. Broodfish usually lose 10 to 30% in body weight due to spawning activity. Weight loss is affected by the number of spawns and length of time fish are held in brood ponds without feeding. Therefore, following removal from spawning ponds, broodfish should be fed sufficiently to replace the weight lost as well as stimulate additional growth. Approximately 3 kg of forage fish per kilogram of bass is required for maintenance, and 5 kg of forage fish will be required for each kilogram of bass for weight gain. Snow[499] fed brood stock with the Oregon Moist Pellet (OMP) through seven seasons without causing discernible differences in fecundity. Broodfish at the San Marcos National Fish Hatchery and Technology Center (SMNFHTC) have been maintained in satisfactory reproductive condition exclusively on formulated feed.

B. Spawning

Largemouth bass normally spawn from spring through early summer at water temperatures of 15 to 24°C.[500,501] The spawning season is generally longer in the southern U.S. than in the north. Fish spawn in central Florida as early as January and may continue until June, though most spawning occurs in March when water temperatures range from 14 to 16°C. Spawning begins as early as February at the SMNFHTC and continues into June. However, peak spawning is from late March through April when water temperatures range from 17 to 22°C. Mature 6-month-old males have been observed constructing nests during October and November and produced milt when stripped.[495] Kramer and Smith[502] reported that bass spawned in Minnesota as early as mid-April and continued until as late as the first of June.

Variation in the timing of spawning is related to water temperature and other environmental variables as well as to the physical condition of adult fish. Many producers believe that strains vary in spawning times, with the Florida strain spawning earlier than northern populations.

Premature spawning must be prevented. Mature male bass in holding ponds will signal the onset of the spawning season by preparing nests.[503,504] Various techniques are used to prevent premature spawning in holding ponds, such as crowding broodfish, manipulating water levels, or running cool water into ponds.[505,506] However, the most effective method results from segregation of the fish by sex.

Holding ponds should be drained and broodfish removed when pond temperatures warm to 10 to 12°C. The sexes are held in separate ponds until water temperature stabilizes at around 15 to 16°C. This normally occurs after the last average frost date. Fish can then be placed into spawning ponds. Spawning may be stimulated if the broodfish are stocked while the ponds are being filled.[507] In general, the best time to stock spawning ponds corresponds with the date when the main spawning period was observed in previous years. If stocking is delayed until water temperatures rise to 20 to 21°C, males will prepare nests, select females, and spawning will normally occur within 24 to 48 hr.

Postponement of the time of spawning may dramatically increase efficiency because the spawning season may be reduced from 6 to 10 weeks to several days. Consequently, fry taken from brood ponds will be more uniform in size and can be distributed or transferred to rearing ponds with minimal losses due to cannibalism and grading stress.

Jackson[508] found that the spawning season can be extended under experimental conditions. Adult fish held in cool water (9.5 to 19.5°C) in New York spawned within 1.5 to 11 days after transfer to 23°C water during the period from June 10 to September 6. In addition, bass have been induced to spawn out-of-season by temperature and photoperiod manipulation.[509,510]

1. Spawning Pond Preparation

Preparation of spawning ponds depends on the objectives and methods of culture. If the fish are to be grown out in the spawning pond (spawning-rearing pond method), fertilization of the spawning pond will be required. Also, rates of stocking broodfish are reduced as compared to the method wherein fry are moved to growout ponds shortly after hatching (fry transfer method).

Ponds should be drained, disk-harrowed, and allowed to dry during the fall and winter. Treatment with an approved preemergent herbicide prior to spring pond flooding will help control rooted vegetation and thus facilitate fry removal.[511] Placement of gravel in selected pond sites (Figure 1) aids in fry collection by inducing nesting and spawning within confined areas.[512] Gravel nests should not be placed in water too deep for wading. Light-colored rocks make the broodfish easier to observe on the nests. Portable spawning mats (e.g., artificial turf or brown carpeting) have been used successfully in a limited number of instances.[513] Fish culturists at the Sand Ridge State Fish Hatchery, Illinois, have used lead-centered cord from gill nets to form circular patterns on the underside of such carpets. In raceways, bass seem to prefer that arrangement since it produces a slight circular depression in the carpet. In ponds, nests should be at least 2 m apart since males exhibit territorial behavior.[493] The number of males stocked should not exceed the number of artificial nests available.

2. Spawning Behavior

The majority of spawning occurs during late afternoon or early morning.[514] Heidinger[494] described the spawning behavior in detail. The male bass prepares a nest by pivoting in a circle, creating a depression in the sediment with a radius of approximately the length of his body. The substrate used may vary, but nests constructed on firm areas will be shallower than those created on soft substrates.

The male will repeatedly leave the nest in search of a ripe female. Changes in color patterns and aggressive behavior are used to entice the female to the nest. The pair

FIGURE 1. Gravel nests (light areas) along the banks of a bass spawning pond.

subsequently slowly circle the nest, side by side, with the male nipping and butting the female. Spawning begins with both fish over the nest and each tilted laterally so their vents are in close proximity. Eggs and sperm are released accompanied by shudders of the body. Spawning takes place over a prolonged period, and multiple spawns by fish of both sexes are common. The male will guard the nest after the female leaves and may entice additional females to lay eggs within the same nest.

In order for the culturist to obtain as many eggs as are available in the brood female population, an excess of males is needed.[515] Bishop[507] recommended 2:1 to 3:1 male to female ratios. Snow et al.[496] suggested stocking adults of the same year class together because they are more likely to spawn at the same time and produce fry of the same size.

No forage fish are placed in spawning ponds because they would compete with fry and perhaps be a source of disease or predation. Broodfish trained to consume pelleted feed can be fed in spawning ponds although feeding is not required.

More precise control of spawning time may be accomplished by means of hormonal injections of ripe fish. Injections with 4000 to 8000 IU/kg human chorionic gonadotropin normally lead to ovulation within 48 hr and will maintain milt production in males.[516] Spawns collected on mats may be moved indoors to hatching troughs, or eggs may be treated with a sodium sulfite solution to dissolve the gelatinous matrix that makes them stick to vegetation.[517] The dissociated eggs may be placed in a Heath vertical incubator for hatching. Collection is scheduled so the eggs will hatch within 24 hr; 2 days after hatching, fry are transferred to holding troughs until stocked. The labor and expense of such intensive methods are not usually warranted, and since large numbers of viable fry are normally produced in spawning ponds, these intensive methods are the exception on most hatcheries.

C. Fry Development

Bass produce from 2000 to 176,000 eggs/kg of body weight.[507,515,518] Fertilized eggs are orange (white in pellet-fed females), adhesive and spherical in shape, with a diameter of 1.5 to 2.5 mm. Egg size increases with size of the producing female. Ovaries contain ova in all stages of development,[519,520] though most of the ripe eggs are released with the initial spawn. Each subsequent spawn normally contains approximately half the eggs in the preceding spawn. A range of about 5000 to 43,000 eggs are deposited in a given nest.[499] Some 78% of the eggs hatch, and 58% of the fry survive to swim-up stage on the average.[521]

Hatching and development times are primarily functions of temperature. At 10, 18, and 28°C, eggs hatch in 317, 55, and 49 hr, respectively.[522,523] Snow et al.[496] found that hatching normally occurs in Alabama 48 to 96 hr after spawning. The transparent fry are about 4 mm TL and disperse in the substrate of the nest. They become pigmented at between 5 and 7 days of age. By that time their yolk sac nutrients are absorbed and gas bladders filled. Fry then rise from the nest in a school and become active predators.[524] Swim-up fry consume cladocerans, rotifers, and copepods.

Although fry abandon the nest 8 to 10 days after hatching, the male parent continues to guard the fry as long as the school remains intact. Fry can be removed to rearing ponds any time after they leave the nest, but to avoid injury to the delicate young fish, it is advisable to wait until they are 15 to 20 mm long before capturing them. That size is normally reached in 10 days to 4 weeks following hatching, depending upon temperature.[496]

When food in spawning ponds becomes scarce, fry must be removed regardless of their size. Up to 1,250,000 fry/ha have been harvested from spawning ponds, but Snow[503] suggested that half that number is a more realistic expectation.

D. Fry Rearing

The aggressive, predaceous nature of the largemouth bass makes it a challenging subject for culture. Fingerlings become piscivorous at lengths between 3.8 and 5.0 cm. The transitional phase from a diet of microcrustaceans and small insects to one of fish is a critical period in the life cycle of the bass. During this natural transition period from one food type to another, fish culturists have been successful in modifying fingerling feeding behavior (i.e., training them to accept formulated feed).

Fry less than 5.0 cm long are usually raised in fertilized ponds. If fingerlings remain in ponds beyond that size or until food becomes scarce, losses from cannibalism will significantly reduce production.[525] Depending on production objectives, three culture methods have been outlined:[503] (1) spawning-rearing pond system, (2) fry transfer system, and (3) intensive system.

1. Spawning-Rearing Pond System

This method is used when the number of required fingerlings is low, commitment of resources and technical expertise are minimal, or available brood stock are of a quality that would not warrant additional expense. Adult fish are stocked into ponds when rising water temperatures stimulate nesting activity. A common stocking rate is 25 to 100 broodfish/ha. The pond is fertilized as necessary to promote development of abundant zooplankton populations. Organic fertilizers (e.g., cottonseed meal, peanut hay, alfalfa, and animal manure), with or without inorganic fertilizers (e.g., superphosphate) have been used to establish and maintain zooplankton blooms. Fry remain in the pond until they reach 2.5 to 5.0 cm, when they are normally harvested for distribution. Depending on temperature, 40 to 65 days are required from spawning to fingerling harvest. The broodfish are returned to holding ponds and maintained on forage fish until the next spawning season. In this method, the number of broodfish used in a pond is the only control exerted over the number of fry produced. Predation on fry by the adults and on fry by other fry may significantly affect the size and number of fingerlings harvested. More than 12,000 fingerlings/ha have been reported.

2. Fry Transfer System

The practice of collecting schools of fry from spawning ponds for transfer to fertilized rearing ponds permits increased production through the employment of more efficient management techniques, but requires more labor and technical expertise. The number and size of broodfish can be controlled to make maximum utilization of spawning ponds. Since fry are harvested within 2 weeks after hatching, spawning ponds should not be fertilized as plankton blooms reduce visibility and interfere with harvest.

a. Fry Collection

Fry are most easily captured while still in schools. Two workers in a boat can maneuver to slide a fine mesh seine (similar to cheesecloth) under the school. Fry are then carefully washed from the seine into a tube of water. More frequently, fry are seined by two people wading. However, this muddies the water making observation of schools difficult. A stone or other weight should be placed in the seine to keep the wind from inverting the net.

Clear, calm water, free of debris and vegetation facilitates fry sightings and removal. Visibility on windy days may be improved by adding vegetable oil to the pond surface to create a slick which will retard wind rippling. Fry are attracted to light, a fact which may be used to advantage during night harvesting.

Fry must be removed from unfertilized spawning ponds before the food supply is exhausted or the schools will break up and the fish begin foraging individually. Single fry can sometimes be observed along shorelines searching for food. Trapping along shorelines with such devices as glass-V traps[526,527] may be an effective harvest method for fry which have abandoned schooling behavior.

Some hatcheries remove fry from spawning ponds in a single operation. Approximately 30 days after spawning, 1.3- to 1.8-cm fry are available for harvest. Ponds are drained and the fry collected in catch basins. Chicken wire screens prevent brood bass from entering the basins. Following fry removal, broodfish are transferred to holding facilities. Ponds may be refilled within 24 hr and brood stock replaced so a second spawn can be produced.

Several methods have been used to estimate fry numbers, including visual inspection, gravimetric techniques, and volume displacement (Figure 2). Swanson[528] estimated numbers of fish stocked on the basis of volume displacement. He placed 250 mℓ of water into a 500-mℓ beaker and added fry until the water level rose to 500 mℓ. Numbers of fry in each 250-mℓ sample were estimated from a count taken of 5 mℓ of fry added to a 50-mℓ graduated cylinder. Any of the techniques mentioned should provide an estimate within about 10% of the actual number of fry stocked.

b. Rearing Pond Fertilization

Preparation of fingerling rearing ponds should begin well in advance of stocking. Ponds should be drained following the summer or fall harvest, exposed to sunlight, and disked.[507] Winter ryegrass is often sown to produce a dense growth of vegetation which provides a source of organic fertilizer upon decay following spring flooding and stimulates prolific zooplankton populations which provide food for fry and fingerlings. Ryegrass cover can also reduce erosion of pond bottoms and levees when the ponds are dry.

Application of Casoron® at the rate of 283 kg/ha immediately prior to pond filling controls submerged and/or rooted vegetation for 5 months at the SMNFHTC facility in Texas. Simazine applied by spraying to exposed pond bottoms at the rate of 22 kg/ha has provided protection from aquatic weeds for up to 6 weeks at the A.E. Woods State Fish Hatchery in the same state.

FIGURE 2.   Estimation of fish numbers can be accomplished by volumetric displacement once the volume of water a known quantity of fry displaces is determined.

Rearing ponds are fertilized following filling. Snow[515] suggested fertilization at the rate of 9 kg/ha nitrogen and 8 kg/ha superphosphate at weekly intervals until an abundant zooplankton is established. Ordinarily, this will require three to four applications. Bowling et al.[529] found that fertilization alone was as effective as ryegrass cover in producing zooplankton blooms. Biweekly applications of fertilizer should be made after the zooplankton bloom becomes established and continued until about 10 days prior to anticipated harvest.[515]

An alternative fertilization scheme[495] involves the application of 168 kg/ha of peanut hay to ponds 2 or 3 weeks before stocking; 10 days following hay application, the ponds are treated with 100 kg/ha of 16-20-0 inorganic fertilizer followed by a second application 10 days later at 50 kg/ha.

An initial application of organic fertilizer (e.g., 1500 kg/ha of alfalfa hay) in the absence of grass cover produces a more immediate zooplankton response than does inorganic fertilizer. Since zooplankters feed on detritus, inorganic fertilizer must be incorporated into plants before becoming available to animals in the form of decaying organic matter. The size and quantity of zooplankters available must be appropriate to the size and quantity of bass fry present; thus very precise timing of fertilization is required.[507]

Zooplankton enumeration and water quality analyses have recently been used to measure pond changes that may affect fish production. Many difficulties remain with respect to pond studies in the areas of zooplankton sampling techniques, water sampling, uncontrolled physical factors, data analysis procedures, and delays in sample analysis. Traditional and new[530] methods have not yet proved reliable. For instance, normally an 80-μm mesh Wisconsin plankton net is used to collect zooplankton; however, many rotifers occur in the range of 50 to 70 μm. Undoubtedly, techniques will be improved to offer new insight into pond dynamics, but presently, fingerling condition may be used as a guide for fertilization schedules. If regular samples indicate fish are not growing or are starving, adequate food is obviously not available, and fertilization or harvest is necessary. Observations of fish condition provide the best present indicator of pond suitability.

c. Rearing Pond Stocking

Snow[515] recommended stocking rearing ponds with 120,000 to 180,000 fry/ha, while McCraren[511] recommended rates up to 240,000 fry/ha, if 2.5 to 5.0 cm fish are desired. In general, stocking density is inversely proportional to desired production size. If objectives involve production of 10.0 cm fish, stocking densities as low as 24,000 fry/ha should be used.[514]

The time required for fry to reach a particular size depends on their size, stocking density, pond productivity, and water temperature. While fingerlings are in the rearing ponds, sound management is necessary for maximum production. That includes proper fertilization, vegetation, predator and disease control, and maintenance of good water quality. Regular sampling of fingerlings provides important information on fish health and growth rates. Fish sampling information, when combined with routine zooplankton sampling, provides instantaneous profiles of fingerling condition and pond productivity. Fingerlings should be moved as soon as growth rates become reduced or size variation becomes acute. The fingerlings may be harvested by seining, trapping and/or pond draining. Procedures used to control vegetation and crayfish must begin before the expected harvest date. Tadpoles may be controlled following harvest.[531]

Fingerlings reach 3.8 to 5.0 cm after 2 to 4 weeks in rearing ponds when water temperatures are in the 21 to 27°C range.[495] White[514] indicated that 3 to 6 weeks are normally required for the production of 2.5- to 5.0-cm fish and 6 to 12 weeks for 5.0- to 10.0-cm fish in Texas hatcheries. Properly managed rearing ponds may yield 30 to 150 kg/ha of 3.8- to 5.0-cm fingerlings. Acceptable survival should be near 90% for 2.5-cm fish, but may be less than 75% for 5.0-cm fish and 50% for fish 10.0-cm or larger.

3. Intensive Culture System

An important contribution to the intensive culture of advanced largemouth bass fingerlings was the development of a standard, nutritionally adequate, economical ration. Although bass fingerlings have been trained to consume ground fish, beef heart, pet food, and a variety of other meat products, production has been limited and unpredictable.[532,533] In many cases the diets were nutritionally inadequate, unpalatable, and the supplies undependable. Snow[534] successfully trained largemouth bass to accept OMP. The diet led to consistently good yields, food conversion, survival, and rates of gain. Nagle[535] reported that a larger percentage of bass accepted feed and grew more rapidly on OMP than on commercial dry diets. A stabilized form of soft diet (Biodiet®) is presently available which requires less stringent storage conditions than the original OMP. Commercial dry diets have been used successfully with larger bass.

a. Training Conditions

Training fingerlings to convert from natural food organisms to a pelleted diet is a critical step in intensive bass production. Fry and small fingerlings do not readily accept prepared feeds. The acceptability of such diets may be a matter of physical characteristics or palatability, or it may be a function of feeding regime. Snow[532] reported 3.0-cm fish are more readily trained on prepared feeds than 1.3-, 1.9-, or 7.0-cm fish. Training fish shorter than 2.5 cm exclusively on formulated feeds has not been particularly successful. Recent research by Willis and Flickinger[536] has shown that bass fry greater than 1.0 cm can be trained to eat carp eggs and can then be gradually weaned to Biodiet®. Fish less than 2.5 cm are best trained using ground fish or carp eggs as a transitional diet from natural to formulated feed.

Training bass to consume prepared feeds is accomplished with fingerlings of 3.8 to 5.0 cm TL, which have been harvested from rearing ponds, cleaned of foreign matter, prophylactically treated for parasites with an approved chemical, and graded to reduce cannibalism and establish uniform size classes. Numbers and weights of fish should be carefully recorded.

Fingerlings are crowded in troughs, circular tanks, or raceways at densities of 4 to 8 kg/m^3 of rearing volume. By crowding the fish together and providing them with their sole source of food, their environment is so altered they are forced to modify their behavior correspondingly to survive. Crowding increases competition for food, which in turn encourages the young bass to learn acceptance of feed pellets. High training densities can be maintained as long as the fish are healthy and water quality remains acceptable.

Westers and Pratt[537] developed hatchery criteria for intensive salmonid production which are applicable to bass culture. A flow rate of four exchanges per hour was suggested, but fewer exchanges are permissible if dissolved oxygen is maintained above 5.0 mg/ℓ and un-ionized ammonia remains less than 0.025 mg/ℓ in the effluent. At SMNFHTC, typical loading rates are approximately 5 kg/m^3 and water is exchanged two to three times hourly.[495]

The ideal water temperature range during training is 25 to 30°C.[503,538] At higher temperatures, bacterial disease (caused by *Flexibacter columnaris*) can lead to serious problems. Incidence of columnaris disease may be reduced by careful handling and grading, training in cool water (e.g., 22°C), and prophylactic treatments with copper sulfate and citric acid (5 to 10 mg/ℓ for 1 hr daily). Bass fingerlings tolerate those concentrations in hard water (250 to 300 mg/ℓ $CaCO_3$), but caution should be used in soft water, as copper toxicity varies inversely with water hardness. If columnaris disease occurs, infected and starving fish should be removed as soon as possible and the remaining fish treated on alternate days with potassium permanganate (2 mg/ℓ for 2 hr), copper sulfate, and citric acid (20 mg/ℓ for 1 hr) until mortality ceases. Fish should be carefully monitored as the treatment is an additional source of stress.

b. Training Methods

Feeding regimen is extremely important because it determines to a great extent the number of fish that will accept artificial feed. Fish may be fed immediately after stocking into training units. The more often the fish are fed, the more often they come in contact with pellets and the greater their opportunity to learn to accept the feed. Feeding eight times daily, at 15% of initial body weight, 7 days a week during training is adequate.

Broadcasting feed around the culture units increases exposure of the fish to the pellets. Research with automatic feeders during the training phase has received some attention, but hand feeding has provided better results to date under production situations.

The recommended size for pellets is 1.0 to 2.5 mm during the initial phases of training. This is increased to 4.8 mm in accordance with the ability of the fish to consume the larger particles. If fingerlings are reluctant to accept feed, it may be advantageous to provide ground fish along with pellets. By reducing the amount of ground fish over time, fingerlings can be weaned to the prepared ration. Williamson[495] successfully trained 80 to 95% of the fish on trial, depending upon strain, when fingerlings weighing 0.25 to 0.35 g/fish were first hand-fed ground fish for 5 days and subsequently weaned to Biodiet® distributed by automatic feeders.

Feed conversion is not a good indicator of training success during the initial training period. The proportion of bass on feed at the conclusion of training is the only valid measure of success. The cost of excessive amounts of feed is low and should be of little concern if the tanks are kept clean to maintain good water quality.

Grading is necessary during the training phase to maintain uniformity among fingerlings, but caution should be used to limit the amount of stress placed on the fish. Diversity in size encourages cannibalism and reduces training success. Cannibals are distinguished by their disproportionately large size or by the presence of tails protruding from their mouths. Emaciated "nonfeeders" should be separated from those that have accepted feed. Grading appears to dissolve any established pecking orders among bass fingerlings and may lead to as high as 75% of the nonfeeders eventually accepting training. At the end of the training period, 65 to 95% of the initially stocked fish should be trained and will have doubled their weight during the training period.[503,511] Before trained fish are transferred to growout chambers, they should be graded into uniform sizes, weighed, and counted to determine training success and the number of fish per unit weight.

Overall, training success depends upon fish density, initial fish size and condition, strain, a nutritionally adequate and palatable diet, and feed of the appropriate size. In addition, an appropriate feeding regimen, grading, elimination of cannibalism, disease prevention and control, and maintenance of good water quality are important aspects which influence training success.

E. Growout Procedures

Following training, fingerlings of the same size are transferred to growout ponds and stocked at 24,000/ha. Alternatively, bass may be reared in raceways. Once on feed, fingerling bass retain the learned behavior indefinitely.[535,539]

Fingerlings should be tempered into new conditions to avoid thermal shock. The fish can be stocked into the catch basin of a pond as it is being filled. Floating mats (1.5 × 1.5 m) made of wooden frames covered with plastic may be placed in ponds for the first few days. Fingerlings concentrate under the mats for protection. Feeding efficiency is increased by providing feed at the mat locations.

During the first 10 days of culture, fingerlings are fed four times daily with 2- to 3-mm pellets at 15% of stocked biomass daily. The amount of feed can be varied depending upon consumption. Feeding rate is gradually reduced during the next week to 10%. That rate is maintained for 30 days after which the fish are fed at the rate of 5% of stocked biomass daily. When the fish reach 10 cm, the diet may be changed to a dry floating or sinking feed of appropriate pellet size and nutritional quality. Nagle[535] found that dry feed reduced labor and feed costs with no reduction in growth rate.

At the Marion, Alabama Fish Hatchery, stocking at 24,000 fish/ha led to bass averaging over 20 cm and nearly 115 g in about 100 days at an average water temperature of 27°C.[540] Survival of at least 80% can be anticipated, with food conversions of about 1.5. McCraren[539] estimated the feed requirements for rearing 20,000 largemouth bass fingerlings to 20 cm at 45 kg of 1.6-mm diameter semimoist pellets, 136 kg of 2.4-mm pellets, and 273 kg of 3.2-mm pellets. An additional 1350 kg of trout feed would be required to complete the feeding program.

F. Handling and Transport

Largemouth bass fry may be shipped in plastic bags which contain a small amount of water and are inflated with oxygen, sealed with rubber bands and placed in insulated containers. Densities of 2000 to 12,000 fry/ℓ have been successfully held for 1 to 4 days under those conditions.[540] Survival to destinations up to 2500 km from the point of debarkation were excellent (87%) and subsequent survival depended on rearing conditions at receiving hatcheries rather than distance shipped. Larger fish may be transported in hauling tanks equipped with aeration devices. Maximum densities for hauling depend on fish size, temperature, and other water quality characteristics as well as on the time of the haul.[514]

Handling, transportation, and changes in water quality are common stresses imposed on largemouth bass during various culture phases. Severe stresses may cause sudden mortalities, but fish often survive temporary stress only to die later of diseases or osmotic dysfunction.[541-543]

Responses of fish to stressors include the general adaptation syndrome described in mammals by Selye[544] in which corticosteroid and catecholamine hormones are released in response to a variety of nonspecific stimuli as part of the responses of an animal in dealing with a stressor. These primary effects initiate secondary responses associated with energy distribution, such as increased blood glucose, decreased liver glycogen, decreased muscle protein, increased heart rate, increased gill blood flow,[543,545,546] as well as diuresis and altered electrolyte and water levels in the blood and tissues.[542,547] Ensuing tertiary responses, including decreased disease resistance, decreased inflammatory reaction and leukocyte migration can also be attributed largely to the action of corticosteroid hormones.[548]

Plasma characteristics of resting largemouth bass may provide an important reference for evaluation of the degree of stress imposed by various management practices. Carmichael et al.[549,550] established normal ranges (mean ± 2 standard deviations) for corticosteroids (0.0 to 3.4 μg/100 mℓ), glucose (13 to 93 mg/dℓ), chloride (84 to 128 meq/ℓ), and osmolality (263 to 326 mOsm/ℓ). Plasma glucose and corticosteroid values increased within 15 min of handling or net confinement and typically returned to normal within 24 hr after the stress is removed.

Depending on the degree of stress, chloride and osmolality changes may not be expressed until after the stress is removed, but may persist for 2 to 3 weeks. Consequently, careful monitoring of fish following any period of stress is important. Although a stress may have been removed, its effects may influence the health of the animal for several weeks.

Transporting fish for any length of time also causes changes in plasma characteristics. Chloride and osmolality values below 70 meq/ℓ and 250 mOsm/ℓ, respectively, are life-threatening and are likely to lead to substantial mortalities.

The use of saline solutions, mild anesthetics, bacteriostats, reduced temperatures, and prophylactic disease treatments apparently reduce stress responses and subsequent mortality. Precapture anesthesia attenuated responses of salmonids to a second stressor[551] and has been recommended for use with striped bass hybrids.[552]

Sodium chloride solutions (0.3 to 1.0%), alone or with some anesthetic (e.g., MS-222 or etomidate), have been widely used in conjunction with hauling, generally with positive results.[552-556] Hauling fish in salt solutions followed by posthaul tempering in salt during a recovery period reduces the stress response. Salt in recovery water lowers the osmotic gradient between plasma and the environment, thereby reducing the energy required for osmoregulation.

Cooled water has been used for fish transport with some success.[557,558] Transportation in chilled water reduces severity of the stress response and osmoregulatory dysfunction similar to that which occurs when salt is added to hauling water. However, the combination of chilling and adding salt provides even greater protection from stress and leads to increased survival.

The most effective method we have used for hauling 200-g bass at a density of 180 g/ℓ for up to 30 hr is (1) treat the fish prophylactically for disease with copper sulfate (10 mg/ℓ) 1 hr daily for 10 days prior to hauling; (2) withhold food from the fish for 72 hr before hauling; (3) anesthetize the fish before capture with MS-222 (50 mg/ℓ) or an equivalent anesthetic; (4) haul the fish in water containing salts similar to those in fish plasma (NaCl, KCl, KH_2PO_4, $MgSO_4$), along with high calcium and bicarbonate (>100 mg/ℓ $CaCO_3$) anesthetize with MS-222 (25 mg/ℓ) and add an antibacterial compound (10 mg/ℓ active furacin), using cool temperature water (e.g., 16°C); (5) temper during recovery in salts similar to the plasma; and (6) use posthauling prophylactic disease treatment (e.g., copper sulfate treatment as described above). Carmichael et al.,[549] found that the described procedures virtually eliminated hauling stress mortality in largemouth bass.

III. SMALLMOUTH BASS

Several culture regimes have been developed for smallmouth bass (*M. dolomieu*) which may also be applicable to spotted bass (*M. punctulatus*). Two methods that have been successful under different hatchery conditions are presented: a raceway method developed by the Tishomingo National Fish Hatchery, Oklahoma, and a pond method adapted from Inslee[559] by the A. E. Woods State Fish Hatchery, Texas.

The technique developed at Tishomingo involves raceway spawning and mechanical hatching. Two advantages of raceway spawning include control over time of spawning and the ability to obtain multiple spawns from the same broodfish. The method also eliminates egg losses when males abandon their nests after a water temperature drop. Smallmouth bass renest and spawn after weather or predators cause males to abandon their nests, though egg numbers in each succeeding spawn are reduced.

The raceway technique involves holding adult smallmouth bass in concrete raceways for 2 to 3 weeks prior to normal spawning time, thus preventing uncontrolled spawning in holding ponds. Wooden nests (40.6 × 10.6 × 10 cm) containing 2.5- to 7.6-cm rocks are placed in the raceways. The nests are constructed from 2.5 × 10.1 cm pine lumber with 0.32-cm hardware cloth covering the bottom. Weighted artificial turf can also be used as spawning substrate in order to reduce labor and facilitate egg handling and/or shipping. Fish will usually spawn on the nests in 24 to 48 hr. Nests are collected from the raceways daily or at least within 48 hr of spawning. Daily observations of nests are made with a glass-bottomed tube made from 15.2-cm PVC pipe.

Mechanical hatching involves placing the entire nest in a catfish hatching facility, directly under the moving paddles. Compressed air forced through weighted Micropor® tubing is bubbled from under the nests. Fresh, heated (23°C) well water is circulated through the hatching troughs. Heated water improves survival to swim-up and hatching occurs in 3 to 5 days. Heated water also reduces losses of eggs from fungus invasion. Standard aluminum rearing troughs (3.7 × 55.9 × 25.4 cm), used without paddles, have led to good hatching success. Nests may be stacked in troughs if necessary. Fry fall through the hardware cloth on the bottom of the nests into the troughs, are collected, and then stocked into rearing troughs at 100,000/trough (about 850 g/trough). Nests and hatching troughs are disinfected with chlorine, rinsed, and air dried prior to reuse.

Heated well water is used in the fry troughs when available, or well water can be mixed with pond water to raise the temperature. Air is supplied throughout the length of the troughs. Circulation provided by the air bubbles helps disperse the fry which have a tendency to aggregate. Vinyl-coated wire mesh (1.9 × 1.9 cm) may be placed on the bottom of the troughs. Fry will tend to occupy squares in the mesh and thus become more evenly distributed. After 3 to 5 days at 20 to 22°C, the fry swim up. Subsequently, they are acclimated to pond water and stocked into rearing ponds for growout. Meat and bone meal can be broadcast into the ponds to supplement the natural zooplankton food source.

The box nest pond culture technique used by Pat Hutson, manager of the A. E. Woods hatchery, offers an alternate method. Broodfish holding ponds are stocked at 225 kg/ha of adult smallmouth bass and 7.2 kg of 5.0- to 7.6-cm goldfish (*Carassius auratus*) are stocked for each kilogram of broodfish. Portable nests filled with rocks are used as spawning substrates. The nests consist of two boxes, one fitting within the other. The outer box (fry collection box) is 53.3 × 53.3 cm, constructed of 2.5 × 15.2 cm cypress lumber, and has a fiberglass window screen bottom. The inner box (nest box) is also constructed of 2.5 × 10.2 cm cypress lumber and is 50.8 × 50.8 cm with a 0.6-cm mesh hardware cloth bottom.

Spawning ponds should have extensive shorelines, silt bottoms with no visible rocks, and gradually sloping levees (3:1 slopes preferred). Pond bottoms are disk-harrowed, bladed, packed, and sprayed with Diuron® to prevent vegetation growth. Box nests are leveled around pond margins at a depth of 45 to 60 cm of water. They are placed 3 to 4 m apart. Rocks ranging from about 5- to 10-cm diameter are placed in the nests.

Ponds are filled 3 to 5 days prior to broodfish stocking. Broodfish are sexed, paired, and stocked when water temperatures are consistently above 16°C. Numbers of males stocked equals the number of nests in the pond. One third more females than males are stocked.

Visual observations of the nests are made daily using an underwater viewing scope constructed from 10.2-cm PVC pipe, 1.5 m long, with a lens from a diving mask attached at one end. After eggs are observed, they are viewed daily until they disappear, indicating hatch. Observations continue until black fry are seen on the rocks. Harvest of the fry is then achieved by slowly raising the entire nest structure to the water surface and gently raising and lowering the nest box. Water currents wash the fry from the rocks, through the hardware cloth and into the collection box. The fry are then rinsed from the collection box into a wash tub for transfer to rearing ponds. Nests are cleaned and reset for subsequent spawns. Ponds are lowered about 1.0 m and refilled with fresh water to initiate spawning activity.

Rearing ponds are stocked with 125,000 to 187,500 fry/ha. Since fry are nonswimming when stocked, they are held in 0.6 × 0.9 × 1.2 m window screen wire boxes within rearing ponds until they swim up. Boxes are placed where fresh water can be added for aeration. When fry are free-swimming they are released. Rearing to 3.8 cm requires 20 to 30 days.

Smallmouth bass are sensitive to handling and hauling stress. In tests designed to measure stress in the species,[560] plasma glucose levels quickly increased after both hauling and handling, indicating that stress had occurred. After the fish were handled and loaded into a hauling unit, plasma sodium and chloride levels dropped. Hauling for 2.5 hr caused further declines in both ions, indicating osmoregulatory dysfunction. Plasma glucose returned to baseline levels within 4 to 5 days after handling, but chloride remained low. The extended hypochloremia resulted from a relatively modest stress; recovery from more harsh hauling conditions may require much longer periods.

IV. OTHER SPECIES

Less research has been conducted on propagation of crappie and the various species of sunfishes and their hybrids than on the basses. Much of the information presented here on sunfishes was compiled by Flickinger and Williamson.[561]

A. Sunfishes

Generally, persons involved with rearing sunfishes have used extensive culture methods. Most literature references pertain to bluegill (*Lepomis macrochirus*) and redear sunfish (*L. microlophus*) culture, but the techniques suggested should apply to most other species and their hybrids.

Bluegills 2 years old or more[562] (18 cm TL, 100 to 150 g) are excellent broodfish.[512] Smaller fish will spawn but production is reduced. Sunfish spawn several times after the water temperature reaches 21°C. Males defend small excavated nesting depressions which are often close together.

Broodfish should be stocked before water temperatures reach 21°C. The number stocked on an area basis determines the size of fingerlings obtained because young and adult sunfish are generally left together in the spawning-rearing ponds. Blosz[563] found that 247 broodfish/ha yielded 370,500 fingerlings/ha which averaged 2.5 cm, while 74 broodfish/ha produced 123,000 fingerlings/ha which averaged 660 fish/kg. Surber[564] reported a harvest of 676,780 small fingerling bluegill/ha from a stock of 198 broodfish/ha. Higgenbotham et al.[562] noted that 375,000 fry can be obtained from 100 broodfish. Huet[512] generalized that stocking 200 to 300 pairs/ha would produce 400,000 fingerlings/ha. Stocking equal numbers of each sex has been recommended,[512,562] but Davis[565] suggested using a ratio of 2:3 (males:females).

Sexes may be sorted according to methods described by McComish.[566] Egg incubation requires less than 5 days at temperatures above 21°C.[512] Since sunfish are fragile and not cannibalistic at small sizes, the young are reared with adults. Spawning-rearing ponds should be fertilized to improve yields. Bluegills accept pelleted feed (this may not be true for all sunfish species and hybrids) and prepared feed can be used to supplement natural forage, leading to an increase in fingerling production[562,565] and the maintenance of broodfish. Pellets should be broadcast throughout the pond. Fines (feed dust) from large pellets can be used as food and fertilizer in spawning-rearing ponds.

Similar culture techniques can be used to produce hybrid sunfishes. Critical to their production is broodfish care and selection. Crossing unpure broodfish does not produce the proper progeny or expected sex ratios. Correct sorting of males and females used in crosses is important as a single error may negate the entire effort. Removing the opercular tabs on males may enhance hybridization success since species recognition, which reduces the incidence of hybridization in nature, is impaired.[567] Young hybrid sunfish can be fed pelleted feed or fines if bluegills are used as one of the parents in the cross.[568]

B. Crappie

Black crappie (*Pomoxis nigromaculatus*) and white crappie (*P. annularis*) behave similarly under culture conditions. In management situations, however, they are dissimilar; 2-year-old crappie are used for spawning, though 1-year-old fish (12.7 cm) will spawn. Few crappie live longer than 4 years.[569] Large crappie feed on fish but will also consume invertebrates.[569] The addition of fathead minnows (*Pimephales promelas*) to broodfish holding ponds may be desirable but is not necessary.

Smeltzer[570] described a method to sex black crappie with nearly 100% accuracy. Portions of that method may also be applicable to white crappie. Basically, selection of fish with deep black coloration should yield adult males. Remaining fish are females and males not in spawning condition. The latter fish are gently stripped. Fish which do not yield eggs are unripe animals of both sexes.

Differences in the urogenital opening are not as distinct in crappie as they are in bluegill.[566] Fish of uncertain sex can be held without food for several days. Any distention of the abdomen above the urogenital opening indicates the presence of developing ovaries. Crappie spawn at water temperatures above 18°C, but males become dark in color at slightly cooler temperatures.

Generalization of stocking rates for broodfish is difficult because production has been variable and in many cases both crappie species have been stocked together and, in some instances, they have been stocked with other species.[570-575] Broodfish densities greater than 247/ha seem to retard production.[570] Yields as high as 164,700 fingerlings/ha have been reported from 125 adults.[570]

Male crappie construct poorly defined nests. Crayfish may reduce production by eating eggs and should be controlled. Egg incubation requires 48 hr at 19°C.[576] Post-spawning adults are normally left in the ponds with fry.

Based on stomach analyses, cannibalism of adults on young is low.[570] Addition of fathead minnows to crappie ponds seems to be detrimental.[570]

One problem in crappie culture involves harvesting. Crappie are extremely susceptible to columnaris disease (also called saddle back disease)[565] and shock.[570] Crappie train well to pelleted feed by the method of Willis and Flickinger.[536] Better training success has been obtained with large (4.3- to 6.6-cm) than small (2.5- to 3.6-cm) fish.[570]

C. Guadalupe Bass

In 1984, a study was conducted at the SMNFHTC to determine if Guadalupe bass (*M. treculi*) could adapt to fish hatchery conditions. Five hormone-injected fish (one female, four males) stocked in a 0.04-ha pond produced about 1200 harvestable fingerlings. Fingerlings (22 to 45 mm TL) were trained to consume pelleted feed with 96.4% success. Growth rates during and following training were similar to other species of *Micropterus*. Cultured Guadalupe bass could be used in stocking programs as a means of supplementing depleted native populations.

V. WATER QUALITY

Environmental requirements of centrarchids vary with species and life stages. Hatching and survival of smallmouth bass are reduced after exposure to dissolved oxygen (DO) concentrations less than 4.2 mg/ℓ at 20°C, and no fish survive for 2 weeks at 2.8 mg/ℓ.[577] However, DO concentrations as low as 2.8 mg/ℓ have no significant effect on survival of largemouth bass eggs.[578] Food conversion efficiency decreases in largemouth bass at DO concentrations below 4 mg/ℓ. Growth is positively correlated with oxygen concentrations up to saturation, but decreases under supersaturated conditions.[579]

Oxygen requirements for growth are inversely related to environmental temperature.[580] The minimum DO levels tolerated by largemouth bass and bluegills are 0.75 and 1.40 mg/ℓ at 35 and 25°C, respectively. Acclimation reduces this tolerance only about 0.15 mg/ℓ.[581] Largemouth bass and bluegill exhibit avoidance of DO concentrations below 3.0 mg/ℓ.[582]

Largemouth bass eggs spawned at 17 to 21°C exhibit good survival (>75%) on exposure to 10 to 24°C, but survival is reduced at temperatures above 24°C. Acclimation at the rate of 0.2°C/hr increases the tolerance range to 10 to 29°C.[583] Largemouth bass fry fed mixed zooplankton and small invertebrates exhibit maximum growth at 27°C, whereas smallmouth bass grow best at 25 to 26°C[584] and bluegill at 30°C.[585] Yearling smallmouth bass grow satisfactorily at temperatures no higher than 32 to 33°C, but tolerate 35°C for short periods.[586]

Bluegill and largemouth bass eggs tolerate salinities of 5 and 11% seawater, respectively. However, fingerling survival is not affected at salinities below 30% seawater.[587] Bluegill acclimated to pH 7.5 exhibit no mortalities due to exposures to pH as low as 4.0,[588] and oxygen consumption is not affected.[589]

LC_{50} values (at 48 and 96 hr) for un-ionized ammonia to bluegill range from 0.5 to 4.6 mg/ℓ.[590,591] Total ammonia nitrogen LC_{50} values for fingerling largemouth bass were 31.4, 17.3, and 15.3 mg/ℓ for 24, 48, and 72 hr, respectively (pH 8, temperature 25°C, alkalinity 232 mg/ℓ, hardness 272 mg/ℓ).[592] Those values correspond to 24, 48, and 72 hr un-ionized ammonia LC_{50} values of 1.69, 0.93, and 0.82 mg/ℓ, respectively. Similar values were reported by Roseboom and Richey.[591]

Although no direct studies on the effects of ammonia on growth of centrarchids have been reported, estimates may be made by comparing the LC_{50} and growth data available for channel catfish.[74,78] Based on the ratio of channel catfish LC_{50} values to the value that inhibits growth and the LC_{50} value for largemouth bass, a continuous exposure to 0.01 mg/ℓ un-ionized ammonia may be expected to significantly reduce growth in largemouth bass. Similar calculations for bluegill[591] suggest that continuous exposure to 0.006 to 0.058 mg/ℓ un-ionized ammonia may significantly reduce growth.

Environmental monovalent anions may reduce toxic effects of nitrite to freshwater fishes.[82] High alkalinity and hardness may also offer some protective effects.[83,593] Both the ionized and un-ionized forms of nitrite are toxic. More nitrous acid is present at acid pH and it apparently more readily enters fish than the ionized form. Bluegill LC_{50} levels for nitrite are 4.0 and 282 mg/ℓ total nitrite at pH 4 and 7.2, respectively (30°C, 48 mg/ℓ total hardness).[594] Values for largemouth bass are 460 mg/ℓ for 96 hr at pH 8 (23°C, 166 to 223 mg/ℓ alkalinity, 205 to 255 mg/ℓ hardness).[595] Bluegill and largemouth bass have greater tolerances to nitrite than most freshwater fishes. Centrarchids apparently prevent nitrite from entering the blood via the gills,[595] whereas all other species investigated actually concentrate nitrite in the blood.

No studies evaluating the effect of nitrite on growth have been reported for centrarchids. However, if channel catfish growth and survival data are used, and the ratio and proportion approach described above is employed, we can estimate that 10.7 mg/ℓ nitrite will inhibit growth in bluegill at pH 7.2 and 17.5 mg/ℓ nitrite will inhibit growth in largemouth bass at pH 8. These estimates should be used with care since other water quality characteristics can have a considerable influence on nitrite toxicity. Also, the nitrite exclusion activities of centrarchids vs. the nitrite concentrating activities of other fishes may cause this estimate to be high.

VI. CONCLUSIONS

Centrarchid culture has evolved slowly but techniques now available allow for successful large-scale rearing of perhaps the most popular warmwater game fish family in North America. Areas still requiring research include nutrition, fry feeding, genetic evaluation, and economics.

Chapter 6

NORTHERN PIKE AND MUSKELLUNGE

Harry Westers

TABLE OF CONTENTS

I. INTRODUCTION

The pike family, Esocidae, represents a highly specialized group of fishes, consisting of only five species belonging to a single genus, *Esox*. Their strong, piscivorous nature presents a special challenge to the fish culturist.

The natural distribution of the northern pike (*Esox lucius*) is circumpolar throughout the temperate and arctic zone of the northern hemisphere, while the muskellunge (*E. masquinongy*) has a rather limited distribution in the eastern U.S. and Canada. The most important countries for esocid production are the U.S., Canada, Germany, Austria, Switzerland, The Netherlands, and to some extent France, Belgium, and Sweden.[596]

The northern pike is primarily cultivated for stocking into natural waters to augment sport and commercial fisheries. Production for direct human consumption, despite the high quality of the flesh, is as yet, limited; but it is reported that in North America, the sale of this species amounted to over $400,000 annually in the 1970s, with the demand for yearling pike far exceeding the supply.[597]

The muskellunge, or muskie, is solely produced to provide a sport fishery, with emphasis on "trophy" fish.[138] In the U.S., artificial propagation of the muskellunge dates back to 1890 in New York and 1899 in Wisconsin.[598]

A recent survey indicates that 23 states and provinces in North America are engaged in muskie management, with all of them employing hatchery stocks in their programs. Over 50% of these states and provinces rely exclusively on hatchery stocks in their management activities.[599]

This chapter describes propagation techniques for the northern pike and muskellunge as well as for their hybrid, the tiger muskellunge. Either sex may be crossed with the opposite sex of the other species to produce these sterile hybrids.

Low hatchability, cannibalism, and severe production variability have been the main barriers to large-scale successful propagation of these species,[512,598] but new culture techniques, including the use of prepared diets, have resulted in significant recent progress.[598,600-604] Because conventional pond rearing often produces erratic results, fish culturists, especially those with experience in salmonid propagation techniques, are searching for ways to adapt esocids to intensive culture methods.

Foundations of this approach were laid in Pennsylvania during the mid-1960s when the first successful rearing of esocids on prepared dry diets was achieved.[603] This represented a major breakthrough, and proved successful on a production scale, first for the hybrid and subsequently, for the northern pike.

Developmental work is continuing, especially in the U.S., to improve upon diets and upon the culture of the muskellunge in general. The economics of muskellunge fingerling production vary greatly, but the intensive method, as applied to hybrids, reflects the cost advantage which can be gained over more extensive culture techniques. Hybrid production costs for intensively cultured 15-cm fingerlings averaged $0.45 per fish as compared with $5.56 per fish when muskie fingerlings were produced in ponds.[597]

II. REPRODUCTION AND GENETICS

A. Reproduction and Spawning
1. Brood Stock Selection

In nearly all cases, wild fish are used as brook stock. They are readily caught on their spawning grounds with either trapnets or pound nets. In some cases, electrofishing is used.

Natural spawning of northern pike normally occurs at temperatures from 5 to 10°C. Males usually arrive on the spawning grounds before the females; usually at the time of ice-out. Muskellunge normally spawn later when the temperature is from 10 to 14°C. Natural hybridization occasionally occurs; however, pike and muskie are not generally compatible because pike are detrimental to muskellunge. The earlier spawning of pike gives them a size advantage over the muskie. The northern pike are thus able to feed upon muskellunge fry.

The use of natural populations for brood stock precludes the development of a selective breeding program. In the case of esocids a selection program favoring domestication for ease of artificial propagation may be desirable. Such a program can only be accomplished by maintaining brood stock in captivity. In North Dakota, muskellunge brood stock were maintained in a 0.3-ha pond on a diet of fathead minnows, perch, and white suckers. Although growth compared favorably with that of wild populations, initial spawning success was poor; 4-year-old females showed egg viability levels of less than 10%.[605]

The difficulty, logistically and economically, of maintaining domestic brood stock of large predatory fish is obvious. Not until adult esocids can be successfully maintained on prepared feeds can significant advances in genetic selection be expected.

2. Spawning and Incubation (Egg Hatching)

Low hatchability has long been a barrier to large-scale production of northern pike and muskellunge. Spawn-taking procedures and egg incubation techniques can both be responsible for poor hatchery success.[598,600]

The use of anesthetics on large broodfish has contributed to significant improvements in fertilization success due to fewer broken eggs. It also results in less stress (damage) to the fish as well as to the person handling the fish. One method used to anesthetize is by spraying a Chlorotone mixture over each gill (1 part trichloro-trimethyl-propanol to 4 parts ethanol). A dose of 1.5 to 2.0 full sprays with a spray bottle of the type used for household window cleaner can bring about deep anesthesia.[600] MS-222 is commonly utilized at a concentration of 100 to 150 mg/ℓ.

Some question the use of anesthetics because they may lull the spawn taker into a state of inattention during which a reflex action by the fish may lead to the partial destruction of a pan of eggs. Instead of anesthetics, some experts advise the use of a sock-like bag to restrain the fish during spawn collection.[606]

Other spawn-taking refinements have been developed, such as the use of a catheter to collect uncontaminated sperm, and employment of a sphygmomanometer (blood pressure cuff) to gently extrude eggs under a controlled pressure of 280 g/cm².[598]

Hypophysation has been used to improve spawning success. Carp pituitary has been successfully employed to accelerate ovulation when injected at levels from 4.5 to 9.0 mg/kg of fish body weight.[598,605,607]

Since males yield sperm in sparse amounts, they have been injected experimentally with progesterone at levels of 10 and 100 mg/kg of body weight. Sperm yield increased two- and threefold, respectively, within 48 hr of injection. There was no adverse effect on fertilization with even the high level of progesterone, and the yield of eyed eggs was 70%.[608]

At times, testes are surgically removed and sperm squeezed from them through cheesecloth.[598] This requires that the brood male must be sacrificed, but leads to increased sperm yield. Normal sperm collection can be expected to provide from 0.2 to 0.3 mℓ from each male. With care, 1 ℓ of eggs can be successfully fertilized (75%) with as little as 0.02 mℓ of sperm.[596]

Sperm will remain viable for several days at 3 to 5°C, but once diluted with water, its activity will cease after 1 or 2 min, depending upon water temperature. At 5°C sperm activity in water may last for up to 2 min.[596]

Changes in incubation techniques have contributed more to improvements in hatching than have changes in spawn-taking techniques.[600] Esocid eggs should be handled very carefully. Fertilized eggs can be transported 1 hr after spawning, but must not be jarred. Eggs are most commonly incubated in some type of jar incubator. It is recommended that eggs not be "rolled" during the first several days of incubation; thereafter, water flow should be increased from about 2.0 ℓ/min to 5.5 ℓ/min for a jar containing 1.0 to 2.5 ℓ of eggs. In Europe, Zoug jars of 6 to 8 ℓ are used to incubate 1 to 5 ℓ of eggs in a flow of 2 to 6 ℓ/min. Eggs are gently rolled during the latter portion of the incubation period.

Fertilization success ranges from 70 to 90%.[512] A good balance of flow must be maintained. Too high an exchange rate will damage the fragile eggs, while if the flow is too low the eggs may clump, depriving them of oxygen.

An alternative method is to place the eggs on small size gravel placed in the bottom of hatching jars. This will prevent rolling during incubation even when a high rate of water exchange is used. Up to 2 ℓ of eggs can be incubated in each 6-ℓ jar. Improvements in percentage of eggs reaching the eyed stage of from 15 to 75% have been reported for northern pike and from 6 to 87% for muskellunge when the eggs are not rolled during incubation.[598] Such dramatic improvements seem convincing, but other investigators have reported good success when the eggs were rolled during incubation.[512,606]

The optimum incubation temperature for esocid eggs ranges from 9 to 13°C, but a range of 5 to 21°C can produce viable fry. Lethal temperatures for northern pike embryos and sac-fry have been reported at 3 and 24°C.[619]

The best results can be expected when incubation occurs under conditions of relatively constant temperature within the optimum range. Dissolved oxygen should be maintained at near saturation. Northern pike eggs incubated at the lower end of the temperature range are reportedly more advanced morphologically than those incubated at higher temperatures. The result is larger than normal fry.[610]

A common practice during incubation is daily treatment with a 1:600 formalin solution for 17 min, or a 1:4000 to 1:6000 formalin solution for 1 hr. Since formalin may harden egg shells, perhaps leading to reduced hatchability, some recommend the use of Diquat at 4 mg/ℓ for 15 min as an alternative to formalin in combating fungus problems.[611]

3. Hatching and Care of Sac-Fry

It requires approximately 120° days (12 days at 10°C) to hatch northern pike eggs and 180° days (18 days at 10°C) to hatch muskellunge eggs. Another 160 to 180° days are required for yolk-sac absorption. The time from initiation of incubation to feeding is from 300 to 360° days, depending somewhat upon water temperature and the species being produced.[512,611]

Since hatching usually encompasses a period of 6 hr or more, it is often advantageous to "force-hatch" the eggs. As soon as first hatching commences, the eggs can be placed in a container and the temperature elevated 5 to 7°C instantly without negative effects. Within a short period of time all of the fry will hatch and can be separated from egg shells, dead embryos, and other debris.

A technique employed in Michigan is to transfer fry immediately after hatching to Heath incubators.[612] The egg baskets must be modified with smaller mesh screens to prevent escapement (Figure 1). The smallest size window screen, approximately 6 × 7 meshes/cm, will suffice. Ideally, the mesh size should be 7 × 7 or, better, 7.5 × 7.5/cm. Retention of fish by the 6 × 7 mesh size can be assured if several coats of paint are sprayed on the screen.

The cover on the incubation basket should be tightly sealed and several layers of artificial turf should be placed in the baskets to keep the fry from piling up and smoth-

FIGURE 1.   Heath incubator trays are modified for hatching esocid eggs.

ering. From 30,000 to 35,000 fry can be placed in a single tray. A normal stack consisting of 16 Heath incubator trays can, therefore, accommodate up to 560,000 fry during yolk-sac absorption. Flow rate should be maintained at 30 to 35 ℓ/min.

The fry are maintained at a temperature of 16 to 17°C and remain in the incubator for 6 to 10 days depending on species. Once or twice during the first 5 days in the incubator, a cleaning operation is undertaken. During that operation the fry are transferred from their incubation tray into a pan and debris is removed.

While in the incubator, treatment of the fry for fungus is not necessary. A savings of about 80% in man hours can be obtained by employing the incubators as compared with rearing sac-fry in troughs or tanks. Sac-fry in Heath trays need a minimum amount of attention. The upwelling water and rapid exchange rates prevent bacterial build-up, a problem associated with the rearing of fry in troughs. The accompanying problem of gill fungus is also avoided in the Heath incubator method. Fry handled in incubators suffer reduced mortality and appear to be healthier than those reared in troughs.

The fry of northern pike possess adhesive papillae on the anterior part of their head just in front of the eyes. These are used by the fish to attach themselves to substrates which are often provided in culture troughs in the form of synthetic cloth, aquatic plants, fir branches, wood, artificial turf, and so forth. This hanging stage lasts about 9 days at 12°C, but clinging is not mandatory to guarantee survival. Fry which are kept in Heath incubator trays during the hanging stage are denied the opportunity to cling but good survival has been experienced. Another interesting phenomenon is the fact that after the hanging stage the fry swim to the surface to fill their swim bladders and from then on adopt a horizontal position in the water.[604]

III. FINGERLING PRODUCTION

A. Extensive or Pond-Rearing Techniques

The production of esocids has historically occurred in extensive culture systems, but the practice is changing to one which utilizes more intensive methods. However, at the present time, most production continues to occur in ponds and it is unlikely that pond culture will ever be entirely abandoned. It is believed that pond culture on natural foods is needed in certain instances for the production of large, high-quality fingerlings required to satisfy certain management strategies.[606]

There are many factors involved in the successful rearing of esocids in ponds. Those which are most difficult to control or which are uncontrollable are related to climate. Climatically induced problems generally are most pronounced during the early culture phase; thus, the critical phase in the culture of northern pike and muskellunge has been identified as that from swimup to 8.0 cm.[606] On the other hand, there are those who feel that the production of fingerlings larger than 10 cm is utopian.[604] The difference is influenced by whether production is viewed from a biological or an economic perspective.

Once esocid fingerlings reach 8.0 to 10.0 cm, enormous numbers of forage fish are required as food. Forage must be continuously present in high concentration. Some culturists recommend a predator:prey ratio of 1:10 to prevent cannibalism. Once the ratio drops below some critical level, heavy losses due to cannibalism will quickly begin to occur and an irreversible trend may become established. It is the need for large numbers of forage of the proper size that makes the production of large esocid fingerlings very expensive. For example, the production of some 11,000 muskie fingerlings averaging 20 cm required 60 million sucker fry and 13,860 kg of minnows.[613]

Extensive pond production of relatively large fingerlings in ponds wherein the forage base is replenished through natural recruitment ranges from only 4 to 100 kg/ha. Semi-intensive pond rearing, wherein forage fish are added at frequent intervals, can routinely yield from 2000 to 4000 kg/ha. In one case, 9000 kg/ha were produced under those conditions.[606] The cost of fingerling production under semi-intensive culture conditions can easily reach $24/kg. Production under those conditions is virtually restricted to rearing of muskies for use in the management of highly valuable trophy sport fisheries. Large northern pike fingerlings could be produced equally as well under semi-intensive conditions if someone were willing to underwrite the cost of such production.

The muskellunge management program for Wisconsin has an annual cost of $190,000, but supports a specialized fishery which has an overall economic value of $3.5 million.[613] The net value of New York's St. Lawrence-Chautaugua-Niagara muskellunge fishery in 1976 was $1.7 million.[614] In 1983 that fishery reportedly contributed $3 to $4 million to the local economy.

Ponds should be properly prepared prior to stocking with esocid fry. Proper fertilization is required for the production of zooplankton blooms. It is recommended that culture ponds be kept dry during the winter. In spring they should be filled and fertilized. Both organic and inorganic fertilizers are commonly used in combination. Rates and frequencies vary depending upon how a given pond responds to treatment. One method which has been used successfully in the mid-western U.S. involves the application of 300 to 400 kg/ha of alfalfa meal or pellets.

The vagaries of spring weather can make production of zooplankton blooms difficult and may also impact upon the newly stocked fish more directly. Muskellunge fry should ideally be stocked when pond temperatures are 16 to 17°C. Should the temperature drop to 8 to 9°C a few days after stocking, as sometimes occurs, the fry may die even in the presence of an abundance of food.[606]

Stocking rates for fry vary greatly. The reported range is from 25,000 to 250,000 fry/ha.[512,615] Stocking rates from 80,000 to 125,000 are recommended.

Feeding activity is temperature dependent in esocids as in other fishes, though optimum temperatures for growth may differ by as much as 6°C depending on the geographical distribution of the different species. Maximum growth for northern pike in Ohio was found to occur at 25°C, while in Ontario the same species grew best at 19°C. Not only does feeding activity increase as optimum temperatures are approached, but success of capture is also greater. For example, northern pike were found to strike at prey six times more often at 20°C than at 13°C and their success in hitting the target was sevenfold as good at the higher temperature.[616]

Cannibalism among esocids can be a major barrier to successful culture. Northern pike fry a few days of age can convert from zooplankton to fish, but they can exist on zooplankton until they reach 10 cm provided the zooplanktonic food is qualitatively and quantitatively correct.[512] In one study it was determined that 2.2-cm northern pike consumed 600 copepods daily and that by the time they reached 5.0 cm each will have ingested from 50,000 to 60,000 planktonic crustaceans. Careful observation of the food supply for young esocids is critical. Evaluation of the zooplankton bloom as many as four to five times daily is not too often. Observations in the early morning are particularly important.[606]

Following the plankton feeding phase, fish must be provided to esocid fingerlings. Fry of the common white sucker (*Catostomus commersonii commersonnii*) are commonly stocked on a daily basis at 10 fish for each esocid for as long as the sucker fry supply lasts. Thereafter common carp (*Cyprinus carpio*) fry or, more reliably, minnows may be fed. A problem with carp fry is that those which are not quickly consumed become unavailable to the esocids since the carp grow more rapidly.

Pond rearing results are often highly variable; survival can range from 0 to over 50%. A consistent survival rate of from 10 to 20% is considered satisfactory for small fingerlings. Once the fish are over 8.0 cm in length, survival to 20 cm can be expected to exceed 60% provided an abundance of forage is available.

B. Intensive Rearing Techniques

In a desire to avoid the variability and unpredictability associated with extensive pond culture on natural foods, fish culturists have sought methods by which esocids can be reared intensively on prepared diets. The first breakthrough came in the late 1960s with the tiger muskellunge.[603] Improvements in rearing techniques as well as diets have since been made. Today the tiger muskellunge, as well as the northern pike, can be reared successfully in tanks on prepared feeds. Successes with the muskellunge have been limited although encouraging progress has been made in Iowa.[617]

There is no doubt in the mind of this author that intensive culture will soon become the prevalent means by which large numbers of esocid fingerlings are produced. Advantages of intensive culture include cost effectiveness. In the Iowa study with muskellunge, pond rearing cost six times more than that for fish reared under intensive conditions. Theoretically, unlimited numbers of fingerlings could be produced under intensive culture since food is not a limiting factor. Production stability and predictability are more completely assured with intensive as compared to extensive culture.

There are, however, certain disadvantages to the intensive culture method. For example, fish reared in close proximity to one another and placed under the stress of a highly artificial environment are more subject to epizootics by various opportunistic as well as obligate pathogenic organisms.

Fry are most vulnerable to mortality during the period when they are trained to accept prepared feed. It is important to begin training with only healthy, vigorous fry. Their health and vigor will relate, in large part, to how the fish were handled during incubation and yolk-sac absorption. Employment of the Heath tray method described above appears to be an excellent means of producing healthy swimup fry.

At swimup, the fry should be transferred to their rearing trough or tank. A rigorous sanitation program should be maintained, even though the intensity of labor required to maintain cleanliness is high, particularly during the early phases of culture.

The fish must be fed to excess to ensure that food is available to them on a continuous basis. Automatic feed dispensers should be utilized. They should be programed to feed every 3 to 5 min over at least 16 hr daily. To protect the fry from fungal infection, daily treatments with formalin are recommended at the rates indicated above. Formalin treatments will also control most protozoan parasites. To control bacterial gill disease, Hyamine, Diquat or Purina-4X can be administered. Chloramine-T at a concentration of 16 mg/ℓ for 30 min has been used successfully therapeutically, or prophylactically by treating routinely once or twice weekly.

Systemic bacterial infections such as Furunculosis and Columnaris can be controlled with Terramycin provided the bacteria are not resistant to the antibiotic. *Pseudomonas* sp. and *Aeromonas hydrophila,* which are able to produce heavy losses in muskellunge fry, have been successfully controlled with UV light treatment of lake and well water supplies during incubation and yolk-sac absorption.[618] In Europe, northern pike eggs are disinfected with Wescodyne (50 mg active I_2 per liter) for 10 min to control Rhabdovirus.[604] Disinfection of muskellunge eggs can be accomplished safely with Povidone-Iodine (1% active I_2) in a 1:100 solution with 10-min exposure.[619]

In general, diseases have not posed a serious problem to intensive esocid culturists. In most instances, epizootics can be successfully prevented or controlled, though it is important that the culturist be constantly alert for signs of disease. The prophylactic treatment approach is recommended, especially during the early phases of culture. Daily formalin treatments during that period are considered essential.

Cannibalism may occur but can be controlled. Proper feeding levels and regimes are the best deterrent to cannibalism, but even when such programs are in force it may be necessary to remove obvious cannibals from the culture system daily, particularly in the case of highly aggressive northern pike. If cannibalism is not detected and circumvented early in its occurrence, high levels of mortality can occur over a short period of time.

Cannibalism can be reduced or prevented during the fry feeding stage by employing the technique of feeding at 3- to 5-min intervals. Once the fish are established on feed, the frequency can be reduced to 10-min intervals. When the fish reach 10 cm or larger, further reduction in feeding intervals to 15 min can be imposed. Feed should be offered for at least 16 hr/day (dawn to dusk), though feeding over 24 hr is not recommended.[601] Two diets developed by personnel with the U.S. Fish and Wildlife Service and used successfully are the W-Series (14-16) coolwater fish diet and the Abernathy Salmon diet (Tables 1 and 2).

Esocids grow much more rapidly than salmonids; thus, it is necessary to adjust feeding levels more frequently than is common for salmonid fishes. Preferred feeding temperatures range from 18 to 22°C, with the lower part of the range most suitable for northern pike. It is best to work with a relatively constant temperature. Most modern hatcheries which are involved with intensive esocid culture are equipped with water temperature control devices. Although temperatures higher than those recommended may produce somewhat more rapid rates of growth, it is advisable to operate below optimum growth temperatures (23 to 24°C) as a means of avoiding problems with pathogens which also grow best in the same temperature range.

The intensive culture method offers the culturist an opportunity to produce large numbers of fingerlings in a limited amount of space, often indoors. Water quality in the culture system can be controlled to a large extent. Production levels per unit area

Table 1
COOLWATER DIET COMPOSITION
(PERCENTAGE OF DIET BASIS)

	Diet		
Ingredient	W14	W15	W16
Herring meal	50.0	50.0	50.0
Blood meal	10.0	10.0	20.0
Shrimp meal	5.0	5.0	5.0
Soy flour	23.0	20.0	10.0
Fish oil (not anchovy)	9.0	12.0	12.0
Vitamin premix	0.6	0.6	0.6
Mineral premix	0.025	0.025	0.025
Ascorbic acid	0.13	0.13	0.13
Choline chloride (50%)	0.225	0.225	0.225
Pellet binder	2.0	2.0	2.0
Proximate composition			
Crude protein	58.2	56.8	61.1
Digestible protein	48.7	47.5	49.8
Crude fat	13.6	16.1	16.0
Metabolizable energy (kcal/kg)	3944	4122	4176

Table 2
ABERNATHY DIET S8-2(84)

Ingredient	Percentage
Herring meal	58.0
Dried whey	10.0
Dried blood or blood meal	10.0
Condensed fish solubles	3.0
Poultry byproduct meal	1.5
Vitamin premix	1.5
Choline chloride	0.58
Ascorbic acid	0.1
Trace mineral mix	0.0005
Pellet binder	2.0
Proximate composition	
Crude protein	48.0
Fish meal protein	40.5
Crude fat	17.0
Moisture	10.0

are high, being expressed in kilograms per cubic meter rather than kilograms per hectare as is the case in ponds. To put the difference in perspective, assume a 1-ha pond, with an average depth of 1 m. Such a pond would have a volume of 10,000 m³. Production might be 400 kg of 20-cm fingerlings, which represents a density of 0.04 kg/m³. In intensive culture a rearing density of 32 kg/m³ is not uncommon. Thus, the ratio of extensive to intensive culture production can be 1:800 or more in terms of space. Thus, production from one 60-m³ raceway could equal that from 4 to 6 ha of ponds.

The following is a description of the rearing program for 200,000, 18-cm fingerling esocids as practiced in a Michigan fish hatchery. Initially, swimup fry numbering well in excess of the target figure are placed directly from Heath incubators into two rearing tanks, each with a volume of 6 m³. Water temperature is maintained at 18°C. Each tank should receive approximately 200,000 fish.

The tanks are fitted with automatic feeders which cover about 90% of the water surface with food when activated. A feeding regime of once every 5 min is initiated with a starter diet of 0.4- to 0.6-mm particle size. Daily or twice daily cleaning of the tanks is undertaken to prevent excessive bacterial build-up. A daily treatment of 1:4000 formalin for 1 hr, flow-through, is employed. Frequent observations are made to determine fry condition, whether they are actively feeding, and whether cannibalism is occurring.

Once the training period has been completed (10 to 15 days at 18°C), the feeding frequency is reduced to every 10 min. Daily formalin treatments are continued and food particle size is increased to 0.6 to 1.0 mm.

When the fish appear to be vigorous and healthy and are actively feeding, a complete inventory is obtained. Numbers are then reduced, if necessary, to 240,000, a level 20% above the target figure. The fish are redistributed among sufficient tanks to complete the tank-rearing phase. The tanks should be equipped with baffles to aid in self-cleaning.[620]

At 18°C esocid fingerlings are expected to grow at least 0.2 cm daily. A hatchery constant of 90 is recommended.[621] The percent body weight (BW) to feed is calculated by the formula HC/L, where HC is the hatchery constant (90) and L represents the length of the fish in cm. Thus, a 10-cm fish should be fed at 9% of body weight daily.

As the fish grow, feed particle size is increased, as is flow rate through the rearing tanks. Ultimately, a maximum flow rate of 400 ℓ/min is reached. That flow rate provides one water exchange every 15 min. If that much water is not available, maximum loading of the tank is reduced. Maximum loading (kg of fish/ℓ/min) can be determined on the basis of the metabolic characteristics of the species, in particular its oxygen requirement.[621] Data indicate that tiger muskellunge consume 110 g of dissolved oxygen (DO) per kilogram of dry feed offered over each 16-hr period. Thus, the maximum loading can be expressed as:

$$\ell/\text{min/kg food} = 110/\text{available DO}$$

The available DO is defined as the level in the incoming water less that selected for the effluent level. If the incoming water contains 8.0 mg/ℓ and the culturist wishes to maintain 5.0 mg/ℓ in the outflow, 3.0 mg/ℓ are available to the fish. Thus, the liters per minute flow required for each kilogram of feed offered is 110/3 = 36.7 ℓ/min. Translating that into kg fish/ℓ/min, the loading equation becomes:

$$\text{kg fish}/\ell/\text{min} = \text{available DO}/(1.1 \times \%\text{BW})$$

and since %BW = 90/8 = 11.25, in this case:

$$\text{kg fish}/\ell/\text{min} = 3/(1.1 \times 11.25) = 0.24$$

A rearing tank receiving 400 ℓ/min can support a maximum of 400 × 0.24 kg = 97 kg of 8.0-cm fingerlings. If the condition factor (K) of the fingerlings is equal to 0.005 (it may range from 0.0042 to 0.0055), the weight of an 8.0-cm fingerling is 0.005 × 8³ = 2.56 g. At maximum loading, a tank receiving 400 ℓ/min can accommodate 97,000 g/2.56 g = 38,000 fish of 8.0 cm.

Redistribution of the fish after initial stocking is required well before they reach 8.0 cm if overloading of the tanks is to be avoided. The best procedure is to redistribute fish after the first complete inventory. In Michigan, final rearing from 8.0 to 18 cm takes place in outdoor raceways, so the initial redistribution is 40,000 fish per tank. Subsequent redistribution is then unnecessary prior to stocking the fish into outdoor raceways.

The raceways have rearing volumes of 60 m³ and maximum inflow rates of 4000 ℓ/ min. Final rearing to 18 cm leads to fish of 29.16 g average (assuming a K of 0.005, the calculation is $0.005 \times 18^3 = 29.16$). The percent body weight to feed for those fish is $90/18 = 5\%$. Maximum loading is $3/(1.1 \times 5.0) = 0.55$ kg fish/ℓ/min. With a maximum flow of 4000 ℓ/min, the maximum weight of 18-cm fish that can be supported in each raceway is $4000 \times 0.55 = 2200$ kg. The maximum number of 18-cm fish which can be supported by each raceway is 2,200,000 g/29.16 g = 75,446 fish.

To produce 200,000 fingerlings, only three such raceways are required. The maximum weight attained, 2200 kg, will result in a rearing density of 37 kg/m³. In Michigan, we consider a rearing density of up to 60 kg/m³ an attainable goal.[612]

From the above description it is obvious that intensive culture of esocids parallels that for salmonids. There is a need to determine more accurately the nutritional requirements of esocids so diets can be improved. Also, esocids have not been domesticated, so genetic selection holds potential as a means of improving upon the results obtained to date. Rearing techniques, especially the difficult and labor intensive early rearing phase, must be improved. The quality of intensively reared fingerlings remains to be fully evaluated, particularly with respect to their utility in management situations.

Sufficient progress has been made during the past decade that intensive culture of esocids on prepared feeds is practical, both biologically and economically. We can expect the technique to be applied in additional situations and with additional species in the future.

Chapter 7

YELLOW PERCH

Roy C. Heidinger and Terrance B. Kayes

TABLE OF CONTENTS

I. INTRODUCTION

Historically the North American yellow perch (*Perca flavescens*) and the Eurasian perch (*P. fluviatilis*) were considered distinct species. A study by Svetovidov and Dorofeeva[622] concluded that there was a single, circumpolar species with three subspecies. This taxonomic status was accepted by some North American authors,[623,624] but not by others.[2] More recently, Collette and Banarescu[625] found that the predorsal bone in *P. flavescens* extends between the first and second neural spine, while it is anterior of the first neural spine in *P. fluviatilis*. This morphological difference clearly separates the two species. Even though Thorpe[626] concluded that the two species are biologically equivalent, the present review is limited to the yellow perch.

Yellow perch can be distinguished from all other North American members of the family Percidae by the following characteristics: the mouth is large with the maxilla extending at least to the midpoint of the eye; canine teeth are not present, but the preopercle is serrate; the anal fin has two spines and six to eight soft rays. Yellow perch are intermediate-sized percids that seldom exceed 0.5 kg, though individuals exceeding 1.9 kg have been reported.[627]

Originally, yellow perch occurred in eastern North America from Labrador to Georgia and west to the Mississippi River.[625,627] The present established range includes much of the U.S. and most of Canada. On the east coast it extends from Nova Scotia south to Florida, then west to northern Missouri and eastern Kansas, northwest to Montana and north to Great Slave Lake (63° N latitude), then southeast to James Bay, Quebec, and New Brunswick. In the U.S., it has been introduced into nearly all states west and south of its original range. Its extensive range and diversity of habitat is a reflection of the great ecological adaptability of the species.

In North America the yellow perch is valuable as a sport, forage, and commercial food fish. By the early 1900s, various agencies cultured yellow perch to the fingerling stage for stocking. Eggs were obtained from the wild and incubated in jars, screened-bottom floating boxes, or wire baskets suspended in streams.[628-630] Leach[628] reported that in 1927, 15 states stocked 12 million eggs, 194 million fry, and 1.25 million fingerlings from U.S. federal hatcheries. Due to these early stockings the yellow perch became naturalized over a large area, and as a result, the demand for stocking decreased so that by 1983 only 0.5 million eggs, fry, and fingerling yellow perch were distributed by federal hatcheries.

Demand for cultured yellow perch depends upon season, price in relation to competing marine species (such as cod and ocean perch), and the commercial catch. Traditionally, people in Wisconsin consume 75% (9 to 11 million kg) of yellow perch which are obtained by commercial fishermen in the Great Lakes.[631] Nearly all commercially caught yellow perch come from Lake Erie and Green Bay, Wis. Following the record 17 million kg commercial harvest of 1969, the mean ex-vessel price of yellow perch in the U.S. and Canada was only $0.25/kg; in 1974 it had increased to $0.71/kg, and in 1976, to $1.64/kg.[632] During 1983 and 1984, the price ranged from approximately $2.00/kg to $3.50/kg. Unbreaded fillets sell for $6.00 to $13.00/kg wholesale and $8.00 to $17.50/kg retail. As the market demand for yellow perch exceeded the supply, interest in developing an economical culture method for the species increased.[633]

Since the current demand for yellow perch is primarily in the North Central U.S., research has centered on intensive recirculating systems which allow control of water temperature for optimum growth. The design parameters for recirculating water systems in which to rear perch are essentially the same as for other fish with similar metabolic rates.

Even though considerable progress has been made in understanding the basic biology of yellow perch, to date no economical method of commercially producing them has been documented in the scientific literature. Thus, the present review emphasizes the known biological characteristics of the species which are relevant to its culture.

II. SPAWNING

Yellow perch are annual spawners with synchronous oocyte growth during fall through winter, culminating in a spring spawning season of approximately 2 weeks. Just prior to spawning, the gonadosomatic index (GSI) of mature females ranges from 20 to 31%.[634, 637] GSI is calculated by dividing the weight of the gonad by the weight of the fish and expressing the number as a percentage. The GSI of male yellow perch may reach 8 to 15%.[635]

West and Leonard[638] found that males mature at 98 to 165 mm (mean = 108 mm) and females at 140 to 191 mm (mean = 158 mm). Clugston et al.[634] reported that a 92-mm male and a 129-mm female were sexually mature. Relative fecundity ranges from 79 to 223 eggs per gram of female body weight.[638,639] Based on the regression equation of Clugston et al.[634] [log fecundity = $-4.21565 + 3.58816$ log total length (TL)], a 130-mm female would have approximately 3000 eggs and a 250-mm female 109,000 eggs. In the wild, a few males mature after 1 year of age and some females at 2 years,[630,640] while most 3- and all 4-year-old fish are mature.[641]

In the spring, males move to the shoreline first, followed by females. Harrington[642] described the actual spawning behavior for fish in the wild; Hergenrader[643] and Kayes[644] described it for fish held in aquaria and tanks. A female is accompanied by 2 to 25 males as she drags the unique transparent, gelatinous, accordion-folded hollow egg tube through the milt. The egg tube (Figure 1) unfolds from the female like a long concertina. It may be several meters long and 10-cm wide.[645] Access to the eggs by sperm and aeration is partially accomplished by water circulation through holes in the gelatinous matrix of the egg tube to the central canal.[646] It takes several minutes at 14 to 15°C and up to several days at temperatures below 5°C, to extrude the entire egg mass.[644] No protection is given the egg mass or young by either parent.

Spawning has been reported at temperatures ranging from 2.8 to 19.9°C.[647,648] Optimum incubation temperatures from fertilization to hatching are considered to be from 10 to 20°C (Table 1). A temperature rise of approximately 0.5°C/day maximizes survival and hatch. Within the range of 13 to 17°C, 85 to 90% of the eggs can be expected to hatch and 70 to 75% of the fry will reach the swimup stage (Figure 2A). Embryos can be incubated at up to 22°C after the neural keel is formed (Figure 2B). Yellow perch appear to require a cooling period (chill period) for late stage yolk deposition and final maturation of the eggs. Hokanson[648] stated that the minimum chill period is 160 days at approximately 10°C or less. He did not obtain viable eggs when the perch were held at 12°C. In yellow perch obtained from Minnesota waters, the optimum chill period was 185 days at 6°C or lower (Figure 3).

Attempts to change the natural spawning cycle of yellow perch by manipulating temperature and light have met with limited success. Of perch exposed to thermal regimes ranging from 4 to 10°C for 120 to 240 days, those exposed to the higher temperatures spawned only slightly earlier than fish exposed to colder temperatures (Figure 4). Kayes and Calbert[649] were not able to significantly change the time of spawning by taking perch from a lake in the winter, warming them to 9 to 13°C, and increasing the photoperiod to either 13.5 or 18 hr of light. Those authors, as well as Hokanson,[648] postulated that within a certain temperature range the onset of spawning depends more on the intrinsic maturation state of the gonads than on photoperiod or temperature cues.

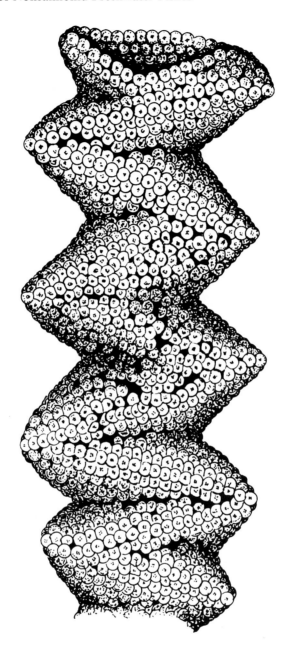

FIGURE 1.   Section of fertilized yellow perch egg mass
that has been extruded from the female.

Hatchability of eggs collected from tank-spawned fish frequently ranges from 40 to
85%. In tanks, females may drop the eggs in the absence of males, or the male, if
present, may not fertilize them.[644] Kayes[644] obtained 80% hatch using the wet and 83%
hatch using the dry stripping method. Fertility was increased to 97% by stripping the
eggs and sperm into a 0.5% sodium chloride solution. Kayes[644] found that the fertility
of eggs exposed to water was reduced from 80%, when fertilization was done within 1
min, to 15% when fertilization was delayed by 5 min. Nearly all the larvae produced
from eggs fertilized after 5 min were deformed. In order to avoid deformities and
increase fertility, he recommended mixing the sperm and eggs within 20 sec after the

Table 1
YELLOW PERCH TEMPERATURE REQUIREMENTS FOR VARIOUS LIFE STAGES

Life stage	Temperature (°C)	
	Optimum	Tolerance range
Maturation	3.9—6.1	<11.1
Spawning	7.8—11.1	2.8—18.9
Cleavage embryo	7.8—12.2	3.9—21.1
Embryo	12.2—16.1	7.7—22.8
Fertilization to hatch	10.0—20.0	6.7—20.0
Hatch to swimup	20.0—23.9	2.8—27.8
Feeding larvae	20.0—23.9[a,b]	10.0—30.0[a]
Juvenile (survival)	23.9—27.8	0.0—33.3
Juvenile (growth)	23.9—27.8	6.1—31.1[b]

[a]   Denotes best estimate based on culture experience.
[b]   Results obtained on excess ration. Restricted rations or mass culture situations will result in a lower growth optimum temperature and upper growth limit which will be dependent on feeding regime.

From Hokanson, K. E. F., in Perch Fingerling Production for Aquaculture, Sea Grant Advisory Rep. 421, Soderberg, R. W., Ed., University of Wisconsin, Madison, 1977, 24.

eggs are stripped. This is usually not a problem because male perch generally have copious amounts of milt.

No attempts to store yellow perch eggs or sperm have been reported in the literature; however, the basic composition of the milt has been determined.[650] Perch milt contains from 1.14 to 3.02 × 10^{10} sperm per milliliter. The pH of the milt was 8.50 and osmotic pressure 316.7 mOsm. The seminal plasma contained 2.64, 0.46, 0.13, and 0.14 mg of sodium, potassium, calcium, and magnesium per gram, respectively. Corresponding cationic concentrations for spermatozoa were 1.81, 1.65, 0.07, and 0.23 mg/g.

The use of hormones to spawn yellow perch is not well documented. In vitro germinal vesicle breakdown did not occur using carp pituitary extracts or mammalian gonadotropins.[651] However, in vitro treatment with various steroids caused germinal vesicle breakdown.[652]

Before water hardening, the diameter of fertilized eggs ranges from 1.6 to 2.1 mm; after hardening, the eggs increase in diameter to 1.7 to 4.5 mm.[645] Hatching time ranges widely, depending on temperature. Eggs require 51, 27, 13, and 6 days to hatch at 5.4, 8.0, 16.0, and 19.7°C, respectively.[645,646,653]

Prior to hatching, the egg mass, which initially has a specific gravity slightly greater than water, loses rigidity and becomes flaccid; a gold iris pigmentation surrounds the melanin in the eye of the larvae; larval movement decreases; and bubbles accumulate in the egg mass, giving it a tendency to rise. This floating tendency can be a problem if the eggs are being incubated in jars.[638] For that reason, Heath incubators have been used to hatch the eggs. They contain the egg mass but do not completely eliminate the floating problem. Hatching occurs within 24 hr after mouth and opercular movements are synchronized in a regular breathing fashion.[646]

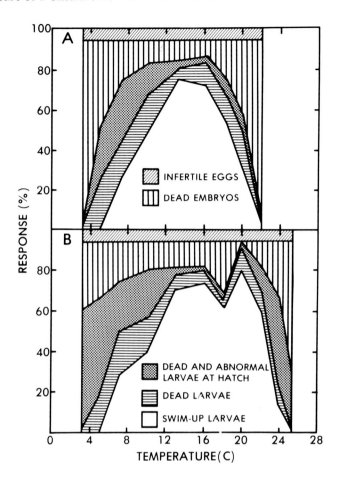

FIGURE 2.   Effect of constant incubation temperatures on percentage total hatch, normal hatch, and swim-up larvae of known aged yellow perch embryos. (A) Embryos incubated at test temperatures from fertilization to larval swim-up stage, (B) embryos incubated at 12°C until neural keel formation, then at test temperatures to larval swim-up stage. (From Hokanson, K. E. F. and Kleiner, C. F., in *The Early Life History of Fish,* Blaxter, J. S., Ed., Springer-Verlag, New York, 1974, 437. With permission.)

III. LARVAL DEVELOPMENT

Newly hatched prolarvae are 4.7 to 6.6 mm TL.[627,645,647] The literature contains conflicting information on when swim bladder inflation occurs. Mansueti[645] indicated it occurs when the prolarvae are approximately 7 mm long, while Hokanson[648] believed that swim bladder inflation takes place during the swimup stage, which at water temperatures above 13°C occurs on the day of hatching. At temperatures below 13°C, swimup occurs within 2 days of hatching. Hokanson[648] further postulated that the prolarvae must fill their swim bladders with air at the water surface. However, Ross et al.[654] found that at temperatures above 20°C the swim bladder inflates in 7 to 10 days after hatching. These differences cannot be attributed to salinity. Mansueti[646] worked with brood fish obtained from a brackish water population but both Hokanson[648] and Ross et al.[654] used brood fish obtained from freshwater.

Prolarvae reach the larval stage at 13 to 14 mm. Although all of the fins are present on fish within that size range, they are not complete until the larvae are 25 to 30 mm

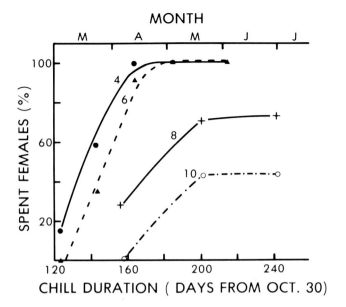

FIGURE 3. Percentages of female yellow perch that spawned during exposure to four chill temperatures (4, 6, 8, and 10°C) of differnt durations (123 to 242 days from October 30). Temperature was increased at the rate of 2°C/week to a maximum of 20°C after termination of the exposure to various chill temperatures. (From Hokanson, K. E. F., *J. Fish Res. Bd. Can.*, 34, 1524, 1977. With permission.)

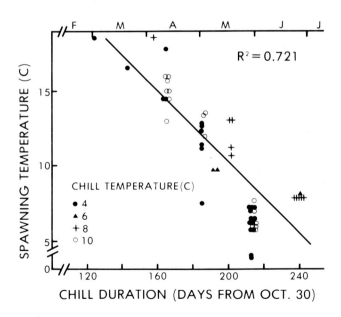

FIGURE 4. Spawning temperature observed at the time of laboratory spawning of yellow perch from Minnesota that had been exposed to four chill temperatures for various periods. (From Hokanson, K. E. F., *J. Fish. Res. Bd. Can.*, 34, 1524, 1977. With permission.)

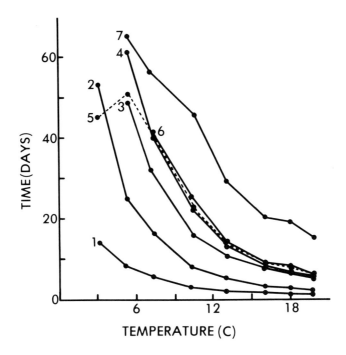

FIGURE 5.   Median times for development to various stages of yellow perch embryos and larvae at various temperatures. (1) neural keel, (2) heart beat, (3) retinal pigmentation, (4) branchial respiration, (5) mass hatch, (6) swimup larvae, (7) mortality of unfed larvae. (From Hokanson, K. E. F. and Kleiner, C. F., in *The Early Life History of Fish*, Blaxter, J. S., Ed., Springer-Verlag, New York, 1974, 437. With permission.)

TL.[645] The median period between swimup and mortality of unfed larvae increases with a decrease in temperature (Figure 5). In one study, mortality occurred in 9 days at 19.8°C and after 21 days at 10.5°C.[647]

Optimum water velocities for rearing yellow perch larvae in intensive culture systems have not been defined. However, Houde[655] determined that perch larvae are better swimmers than walleye larvae for length classes less than 9 mm; swimming ability of the two species is equal between 9 and 15 mm. Velocities that larvae under 9.5 mm can sustain are less than 3 cm/sec. Thus, they are poor swimmers. Larger larvae can sustain current velocities of 3 to 4 body lengths per second for at least an hour.

Optimum temperatures for rearing and feeding larval perch are between 20.0 and 23.9°C (Table 1). This range corresponds closely with their thermal preference.[654,657] Researchers have been relatively unsuccessful in trying to train larval perch less than 18 mm to accept prepared diets, and brine shrimp nauplii do not appear to be an adequate diet for larvae less than 13 mm long.[645] However, yellow perch larvae have been raised experimentally on mixed zooplankton obtained from lakes.[658] According to Hale and Carlson,[658] 250 zooplanktonic organisms per larval yellow perch daily are required to obtain 50% survival during the first 3 weeks of feeding. Those authors recommended feeding at least four times daily and using fry tanks with dark bottoms. Egestion time for larval perch held between 20 and 23°C is less than 1 hr.[656]

The thermal tolerance of successive embryonic and larval stages of yellow perch increases with morphological differentiation.[647] The optimum temperature for feeding and rearing juvenile perch is 23.9 to 27.8°C.[659,660] This is slightly higher than the 20 to 24°C range found by Huh.[661] The upper incipient lethal temperature for juveniles is

29.1°C and for adults 33°C.[648] Deformities occurred at 32°C, and all fish died within 7 days at 34°C. Little growth occurred below 8°C.[660]

West and Leonard[638] were able to train 0.38-g yellow perch to accept prepared feed with 38% survival. Stairs[662] successfully trained an unspecified percentage of 18- to 25-mm yellow perch to accept a prepared ration. Survival is directly related to fingerling size at the beginning of the training period. Fewer than 50% of larvae less than 16 mm can be expected to survive, while 80% of larvae 18 mm long and 98% of those longer than 31 mm have been shown to survive when started on the Spearfish W-7 starter mash and #1 crumble (600 to 800 μm).[663]

IV. GROWTH

Mean growth rate of yellow perch from natural populations in North America, calculated from data presented in Carlander,[664] indicates that the fish reach 7.4, 13.3, 18.0, 20.4, and 22.9 cm, respectively during their first 5 years of life. The current market size for yellow perch is about 20 cm (150 g). Based on experimental rearing trials, Calbert and Huh[665] postulated that 1.0- to 1.5-g yellow perch will reach market size in 9 to 11 months when cultured at 21°C, maintained at 16 hr light and fed 3 to 4% of body weight daily. Those authors estimated that feed conversion ratio would be about 1.5 on a high quality feed. In terms of growth rate and feed conversion in fish larger than 12 g, they found little difference among feeds containing 27, 40, and 50% protein. However, protein-energy ratios were not held constant in their study.

Photoperiod, in addition to temperature, appears to have a significant effect on growth rate in yellow perch.[661,666] Huh et al.[666] found that growth of fish reared at the same temperature, but with a short photoperiod (8 hr light), was only about one third that of perch exposed to longer photoperiods (16 or 20 hr light). Perch of 9 to 17 g reared under 16 hr of light at 22°C and fed at 3% of body weight daily on the Spearfish W-3 diet with 3% added gelatin assimilated 80% of the daily ration and converted 67% of the total food energy into body substance.[661]

In terms of food intake at 25°C, yellow perch appear to be rather tolerant of oxygen levels down to 3.8 mg/ℓ, as well as to diurnally fluctuating oxygen levels.[667] The mean routine oxygen consumption for 1.8-g perch was 0.153, 0.250, and 0.426 mg O_2/hr/g at 15, 20, and 25°C ($\log Y = 1.486 + 0.47x$). Active oxygen consumption was approximately 0.500, 0.540, 0.610, and 1.500 mg O_2/hr/g at 15, 20, 25 and 30°C.[661] In tanks, perch over 50 g are not very active.

Female yellow perch grow considerably faster and reach larger ultimate sizes than male perch.[627,664] In a laboratory study, females commenced growing faster than males at 110 mm (15 g).[668] Since the difference in growth rate expresses itself long before the fish reach marketable size, there has been some interest in reversing the phenotypic sex of genetically male yellow perch with hormones. Starting with 20- to 35-mm fish, treatment for 85 days with estradiol-17B at 15, 30, 60, and 120 mg/kg of diet induced sex reversal of germ cells in most genetic males.[669] No percentage data or information on the growth rate of the fish was provided.

V. CULTURE

Yellow perch fingerlings can be raised in predator-free fertilized rearing ponds for stocking or subsequent grow-out as food fish. Various methods have been used to fertilize ponds, depending upon the location of the culture unit and the availability of inorganic or organic fertilizer. The goal has been to produce a dense bloom of zooplankton without the production of excess vegetation. Few actual fingerling production values occur in the literature, especially under conditions where attempts have been made to maximize production. West and Leonard[638] stocked a 0.23-ha pond with 111,000 fertilized yellow perch eggs in anchored, floating screens (30 cm × 30 cm × 3.8 cm) made of 6-mm mesh. They harvested 35,789 fingerlings averaging 0.38 g; equivalent to 35,789 fish/ha.

Eggs, obtained from the wild, by tank spawning or by stripping can also be incubated in trays such as those found in Heath incubators. After hatching, the fry may be stocked into fertilized rearing ponds.

Fingerlings can be harvested from rearing ponds by seining, drawing the pond down to the harvest basin or, since young perch are positively phototaxic,[670] with the aid of a light and lift net. Manci et al.[671] found that perch from 8 to 50 mm are attracted to light. They successfully removed 61% of the fingerlings (23,000) from a 0.08-ha pond with a 132 cm × 366 cm rectangular lift net constructed of 3-mm white knotless nylon mesh.

As in the case with other fish species, it is not economically feasible to produce yellow perch commercially to food fish size on natural foods. Thus, it is necessary to train fingerlings to accept prepared diets. After fingerling yellow perch reach 18 mm in length, 80% of them can be trained to accept prepared diets. The training procedure for perch is essentially the same as that used for largemouth bass[494] or striped bass.[672] Significant components of the training procedure include concentrating the fingerlings, removing the natural food source, elevating the temperature to ensure an aggressive feeding response, feeding frequently, and grading fingerlings to reduce cannibalism. As in the case with largemouth bass and striped bass, an initial feed with a soft texture (such as the Oregon Moist Pellet) appears to improve training success. After the fish are trained, a hard pellet may be substituted for the soft diet. Fingerling yellow perch can consume 0.84- to 1.19-mm diameter pellets.[673]

Trained fingerlings can be placed in ponds or tanks and raised to marketable size. In the North Central states, 2 to 3 years are required for the fish to reach marketable size in ponds on natural foods. A bioenergetics model developed by Kitchell et al.[674] predicts that yellow perch under ambient (Michigan-Wisconsin) temperature conditions, including winter, should reach marketable size in 15 to 18 months if food is not limiting. In recirculating water systems where temperature and light can be controlled, the rearing period might be reduced by as much as 50%.

Large production studies which include economics are lacking for yellow perch reared in ponds. West and Leonard[638] estimated that in 1977 dollars it would cost $2.31/kg to produce 23,700 kg of yellow perch in a tank system. Kocurek[675] estimated the cost at $8.65/kg. Based on the increase in the wholesale price index, these costs in 1985 dollars would approximately double. The economics are not favorable, since the 1984 market value of perch ranged from $1.98/kg to $3.52/kg. Either the selling price would have to increase significantly or production costs must be reduced.

One problem associated with reducing production costs is that genetically the yellow perch is a relatively small fish, and even though it can be marketed at a small size, it may go through its rapid growth phase before it reaches the desired 150 g. Further, even if one assumes 100% trainable fingerlings and 100% survival, each kilogram of perch represents approximately 6.5 fingerlings. Thus, the cost of fingerlings accounts for a considerable percentage of the selling price per kilogram of fish produced.

This argument carries through to the restaurant menu. By hand filleting wild fish and leaving the belly tissue intact (butterfly fillet), dress-out yields that average 45% are obtained; very diligent grading and constant equipment adjustment, machine processing can yield dress-outs of 42%;[676] however, 37 to 40% is more realistic. The dress-out weight of cultured fish is approximately 5% higher than that of wild fish.[665] Thus, if the mean weight of a serving is 123 g and the mean weight of a bufferfly fillet is 64 g, then two fish are needed per serving. Assuming a training success of 80% and 80%

survival of trained fish, the cost of producing three fingerlings must be reflected in each serving.

VI. CONCLUSION

State and federal fish hatcheries can produce yellow perch fry and fingerlings for use in fish management. The economic feasibility of commercially culturing yellow perch for the food fish market has not been documented in the scientific literature. It is doubtful that the demand for yellow perch will elevate the price sufficiently to make it economically feasible to produce the fish in recirculating water systems; however, neither the market, nor the price that the fish would command outside the states bordering the Great Lakes has been tested. Also, the feasibility and costs associated with rearing yellow perch to market size in ponds or cages under various ambient temperature conditions have not been examined.

Chapter 8

WALLEYE

John G. Nickum

TABLE OF CONTENTS

I. INTRODUCTION

In North America, two fishes within the family Percidae have been the focus of a considerable amount of research by fish culturists. One of those, the yellow perch (*Perca flavescens*), was considered in Chapter 7. The other, *Stizostedion vitreum*, is considered here. Two subspecies of *S. vitreum*, *S. v. vitreum* (walleye) and *S. v. glaucum* (blue pike), have been recognized,[2] but aquaculture attention has been directed only on the walleye. Culture interest has also developed for the sauger (*S. canadense*) and saugeye (hybrid between *S. canadense* and *S. vitreum vitreum*), though relatively little information is available on those fishes. Other common names for the walleye are yellow walleye, pickerel, yellow pickerel, pikeperch, yellow pikeperch, walleye pike, pike, and yellow pike.[677]

Original distribution of the walleye was in the fresh waters of Canada from Quebec eastward. In the U.S. walleye were native from the St. Lawrence River south to the Gulf coat of Alabama, but not east of the Appalachians. The western extent of their range in the U.S. was east of a line from the western border of Alabama through the northeastern corner of Montana, essentially following the northern boreal and central and southern hardwood forests.[627] While the fish is largely restricted to freshwater lakes and streams, there have been isolated reports of walleye occurring in brackish water.[627]

The walleye is one of the most valuable and sought-after fishes in North America, both as a commercial species and as a game fish. Commercial production from Lake Erie alone totaled nearly 14,000 tons in the late 1950s.[678] States such as Minnesota, Wisconsin, and Michigan that feature walleye fishing consistently rank near the top in both resident and nonresident angling licenses sold. Walleye have been introduced into lakes throughout much of the U.S. Most culture has been by state and federal hatcheries for the purpose of stocking recreational fishing waters. Much of the emphasis on walleye production will continue in that vein; however the popularity of walleye with consumers implies that a foodfish market could also be established if economical culture of edible fish becomes possible.

Techniques for taking and incubating walleye eggs stripped from wild-captured fish are well established. Although the origins of that aspect of walleye culture have not been recorded with certainty, it is known that fry have been used in stocking programs for over 100 years.[679] Cobb[680] described walleye propagation in Minnesota over 60 years ago, and Nevin[681] discussed techniques for "hatching the wall-eyed pike" at the 16th annual meeting of the American Fisheries Society in 1887. The general procedures employed by early culturists have not changed substantially in the intervening years.

Management needs in some areas call for the production of fingerling walleye of large sizes. Until recently, the only system available for rearing walleyes to fingerling size was pond culture. However, neither large numbers, nor even predictable production can be obtained from pond culture. In spite of somewhat erratic pond production, most states continue to produce walleyes in that type of system.

The demand for more and larger walleyes for sport fishery management, as well as the potential for commercial fish farming have produced strong interest in intensive walleye culture. That methodology holds potential for rearing walleyes to any desired size, though certain breakthroughs in technology will be required before economical intensive culture becomes a reality. Basic techniques for intensive culture were described by Nickum[682] in 1978 and experimental work has since continued. The Pennsylvania Fish Commission has used intensive culture procedures for practical production of walleye fingerlings up to 10 cm and the Iowa Conservation Commission plans to do so in 1985.

Since successful rearing of first-feeding fry under intensive culture conditions has not been economically accomplished at production levels, all intensive walleye culture has been based on pond rearing until the fish reach at least 2.5 cm. Thereafter, the fish may be offered formulated feeds and reared under hatchery conditions.

Although the prediction that intensive walleye culture would be well established by 1988[682] now seems overly optimistic, interest in walleye culture has remained high. Predictable advances during the next few years can be expected to bring large-scale walleye culture from the level of fry production only to that wherein fish of various sizes will be routinely produced.

II. LIFE HISTORY

A. Environmental Conditions

Walleye juveniles and adults can be found in lakes above the thermocline. They often occur over relatively deep waters during summer and move inshore in early autumn as water temperatures begin to fall.[683,684] The species can be found over a temperature range of 0 to 30°C, though the preferred range is about 20 to 23°C.[685]

Adult walleye are negatively phototaxic[686] and may spend most of the daylight period in contact with the substrate or hiding under various objects in the water.[687] Walleyes feed primarily at night in shallow water[687,688] and migrate to shoal areas diurnally.

Walleye are able to tolerate a wide range of turbidity[687] and often occur in highly colored lakes which are rich in humic acids.[678] They tolerate dissolved oxygen (DO) concentrations as low as 2.0 mg/ℓ in the laboratory,[689] though in nature walleyes are generally found at DO levels above 3.0 mg/ℓ.[690] Walleye typically occur over a pH range of 6 to 9.[689]

B. Reproduction

Maturity in walleyes depends on water temperature and may also be a function of food availability.[678] In general, males mature at 2 to 4 years and females between 3 to 6 years of age. Sizes at maturity are >279 mm total length (TL) for males and 356 to 432 mm TL for females.[627]

Spawning behavior in nature may vary considerably and has been described by Eschmeyer[691] for lakes, Ellis and Giles[692] for streams, and Priegel[693] for marshes. Group spawning appears to be a common phenomenon in walleye.

Spawning occurs in shallow water — often in <1 m of water in lakes — and over various bottom types.[691] Milt and eggs are spawned into the water column, where fertilization subsequently occurs.[677]

The number of eggs per unit body weight of female is relatively constant within a given population of fish, but has been found to vary from 28,000/kg[694,695] to over 120,000/kg.[696,697] Values of approximately 60,000 eggs per kilogram body weight are typical throughout most of the walleye's range.

The spawning season may begin as early as January,[698] and in some regions may not be completed until June,[627] depending upon latitude. In some northern parts of their range, walleye may not spawn during years when temperatures are unfavorably cold.[627] In general, spawning begins shortly after the ice breaks up in spring.[677] A so-called chill period may be required to induce spawning as walleyes not exposed to certain minimum temperatures, below about 10°C in at least some instances,[677] will not reproduce.

Walleye eggs are adhesive when spawned. The adhesiveness lasts for an hour or more[699,700] during which the eggs become water-hardened.[699] Development rate is dependent upon temperature. As reviewed by Colby et al.,[677] 10 days are required for hatching at 12.8°C, while the eggs will hatch in 4 days at 23.9°C. Best hatching success rates have been achieved at intermediate temperatures, though incubation at the low end of the above range leads to a higher percentage hatch than at the high end. Koenst and Smith[701] obtained greatest hatching success at 6.0°C, though no fry survived the period of yolk-sac absorption. Those authors obtained best survival from fertilization through yolk-sac absorption at temperatures between 8.9 and 12°C. A great deal of variability in walleye egg viability in nature has been reported in the literature. For example, instances of hatch rates as low as 3.4%[702] and as high as 100%[703] have been reported.

C. Natural Food

Walleye fingerlings and adults are highly piscivorous except during late spring and early summer when invertebrates may be important foods for fingerlings. Several authors have demonstrated that young-of-the-year yellow perch are a primary prey species in instances where that species is available. Various other fishes are taken when the preferred prey are not present. The above information is reviewed in detail by Colby et al.[677]

Limited information is available on first-feeding walleye fry, but reports indicate that fry up to 7 to 8 mm consume phytoplankton exclusively in Lake Erie.[704,705] In rearing ponds, various types of zooplankters make up the food of walleyes in the 5 to 9 mm TL range (reviewed by Colby et al.[677]). Selectivity of zooplankton by walleye fry was studied in the laboratory by Raisanen and Applegate.[706]

III. CULTURE

A. Spawning and Hatching

To date, attempts to domesticate walleye by maintaining successive generations in captivity have not been carried beyond the second generation, though researchers at the Fisheries Research Laboratory, Southern Illinois University are rearing a group of walleyes spawned by workers at the Iowa Cooperative Fishery Research Unit with the intent of initiating a long-term domestication program. Bandow[707] compared growth of the offspring from pond-reared walleye with those obtained from wild-captured fish, but did not show enhanced growth in fingerlings produced from pond-reared adults.

Spawners are obtained most often from trap-nets set in selected locations in streams or lakes.[677] Spawning generally occurs at water temperatures from 6.7 to 8.9°C[627] so temperature recommendations for setting nets to trap walleye brood stock (7.2 to 10.0°C) which date back to the first quarter of the present century[708] are generally appropriate.

Collections of spawners are made at sites known to be natural spawning grounds. Walleyes typically spawn in relatively shallow water over substrates of gravel and/or rubble. Water as deep as 4.5 m and substrates of sand[691] have been reported as spawning sites, however. The nature of the bottom and water movement and exchange must be sufficient to provide adequate oxygen for the developing embryos; thus, spawning seldom occurs over silt or muddy bottoms.

Males generally arrive over the spawning grounds before females and may remain for several days after the females have left. Even collections made at the peak of spawning activity are often dominated by males.

Broodfish are commonly stripped at the site of capture, though if unripe, they may be held in pens or tanks until sexual maturity is reached.[677] Many hatcheries have developed facilities in which to hold recently captured fish so that spawning at the peak of development can be achieved. Holding broodfish also ensures that sufficient numbers of both sexes are available.

Experienced walleye culturists can identify mature females on the basis of general body shape, particularly the extended, somewhat softer belly as compared to males.

However, no absolutely reliable external characteristics for identification of the sexes has been established.

No special procedures beyond normal care to minimize stress are used in the transport and holding of walleye spawners. While most culturists do not employ sex hormone injections and no widely accepted methodology has been developed, ovulation in wild-captured prespawning adult females can be induced by injections of acid-dried carp pituitary[700,709] and human chorionic gonadotropin (HCG).[709] Injections of 7.7 to 22 mg/kg of carp pituitary at 72-hr intervals or HCG at the rate of 152 IU/kg are effective in inducing ovulation.[710] Pond-reared walleyes can also be induced with HCG.[710] If attempts are made to force ovulation with hormones too far in advance of natural spawning, the eggs will flow freely but will not be viable.[711]

Eggs and sperm may be mixed in a dry pan (dry method) or the pan may be dipped in water and shaken relatively free of water film before eggs are added (wet method). The milt of two or more males is added to the pan of eggs in either method and the mixture is stirred. Olson[712] suggested that maximum fertilization could be achieved by mixing the eggs with water before the milt is added.

To prevent the eggs from clumping due to their adhesiveness, compounds such as starch and bentonite[565,713] can be added to the eggs during the water-hardening period. Fuller's earth may be added at 35 to 40 g/ℓ of water in a slurry that is poured over the eggs and stirred for approximately 5 min, after which it is rinsed free. Tannic acid is also popular with some fish culturists.[714,715] Waltemyer[713] recommended adding tannic acid after fertilization and an initial rinse to remove excess milt. Reduced adhesiveness was observed by Dumas and Brand,[716] who added a solution of 130 mg/ℓ tannic acid to fertilized eggs. More recently, Colesante and Youmans[717] recommended exposure of walleye eggs for 2 to 3 min to a 400 mg/ℓ tannic acid solution.

Once the eggs become water-hardened (1 to 2 hr), they are rinsed and placed in hatching jars such as Downing jars.[717] Water flows (3 to 6 ℓ/min) are adjusted to produce a gentle "rolling" movement of eggs throughout the 3.8-ℓ jars. Hatchery workers in Pennsylvania add a layer of gravel to the bottom of each jar, which produces a diffuse upwelling of water with no rolling of the eggs. Their hatching rates are similar to those obtained in other hatcheries. The volume of eggs added to each jar is largely a matter of personal preference; however, jars are seldom filled to over 75% of total volume. The number of eggs present in each jar can be estimated by volumetric procedures, though walleye eggs seem to vary slightly in size from stock to stock and even as a function of female size and condition; therefore, counts of known volumes from each lot of eggs should be made if accurate estimates of egg numbers are desired. Values of 30,000 to 35,000 eggs per liter are typical.

The length of incubation time required for hatching is temperature dependent, ranging from approximately 3 weeks at 10°C to 1 week at 20°C.[701] Many walleye hatcheries use surface water supplies with variable temperatures. Other local conditions may also affect the length of time required for hatching; therefore, no fully reliable guide can be offered. Colby et al.[677] included information derived from the literature on incubation times, but did not develop standards based on temperature units or other universal units. Hatching success rates of 55 to 70% are typical. Methods used to estimate egg and fry numbers have been less accurate than desirable, so while hatching rates of 95% have been reported, they have not been verified.

Walleye eggs do not require substantial care during incubation. It is not standard practice to remove dead eggs from hatching jars, though clumps of dead eggs held together by the hyphae of fungi should be manually removed. Some culturists apply routine prophylactic treatments with various fungicides to reduce egg loss, for example, workers in Pennsylvania hatcheries employ a 1:600 formalin bath for 17 min daily as a means of controlling fungus on eggs. However no specific procedure or treatment can be recommended for general use. Jars should be observed several times daily to monitor water flow rates, egg condition, and larval development.

Larvae may be transferred from the hatching jars to holding tanks. In some hatcheries, they are allowed to swim out of the jars into holding tanks. Larvae may be transported immediately after hatching and are often stocked into rearing ponds or lakes at swimup. Some managers prefer to wait 1 to 3 days following hatching before stocking walleye. Larval walleyes must obtain adequate food within 3 to 5 days posthatch or they will reach a point in the starvation process which cannot be reversed even in the presence of abundant food supplies.[718] The time required for yolk-sac absorption and initiation of feeding is related to temperature; for example, the critical time for onset of feeding is 3 days at 20°C.

B. Pond Culture Techniques

Culture of walleyes in ponds is dependent upon the production of adequate supplies of food for the young fish. Walleye fry generally feed on zooplankton;[719,721] however diatoms have also been found as major stomach content items in fry recovered from Lake Erie.[722,723] Walleye often become piscivorous when they reach lengths of 6 to 8 cm, though Walker and Applegate[724] found that they would remain planktivorous to 9 cm or larger if abundant supplies of zooplankton were available.

Management of ponds to produce and maintain large zooplankton populations is still a mixture of art and science. Many biotic and abiotic variables interact to affect plankton community dynamics, so fish culturists must be able to analyze, interpret, and then manipulate conditions in their ponds on a continuing basis. Dobie[721,725] developed fertilization techniques that substantially increased walleye production in drainable Minnesota ponds. His techniques were based primarily on maintaining more than 4% organic matter in pond bottom soils. He recommended the use of organic fertilizers such as sheep and otehr barnyard manures.

More recently Geiger[530,726] has developed systems using liquid inorganic fertilizers and inoculation of seed stocks of desired zooplankters as means of increasing fingerling production in striped bass ponds. That methodology has not as yet been fully tested in walleye production ponds. Currently, culturists responsible for pond-rearing walleyes use various combinations of inorganic and organic fertilizers in a rather subjective manner to manipulate plankton densities. Each set of ponds, and even individual ponds within a set, seems to require independent management. The techniques employed for individual ponds are often dependent upon the experience of the culturist. Even after many years of experience, the culturist is often unable to predict fingerling production from year to year or from pond to pond.

Cannibalism is a major source of fry loss in walleye ponds. Problems with cannibalism increase as zooplankton populations diminish and as differential growth creates substantial variation in fish size within ponds. Harvest before the fingerlings reach 6 to 8 cm can increase survival and size uniformity, though fish in that size range may not be desired by the user.

Walleye fry typically reach 5 to 6 cm in 6 weeks or less, depending upon stocking density and food supply in the pond. Lengths of 20 to 25 cm can be obtained in 12 weeks; however, the numbers of fish harvested of that size are generally quite low.

Ponds are usually stocked with at least 125,000 fry/ha. Stocking rates as high as 375,000/ha have been used, but growth rates and survival tend to be poor in the latter instance. Survival to harvest can be in excess of 50% when pond fertility is high and the fish are harvested before they reach 5 cm. Survival to harvest of fish 10 cm or larger can be less than 1%, particularly when pond fertility is poor. Cheshire and Steel[727] documented this negative logrithmic relationship between survival and length at harvest of walleye in ponds.

Harvest of walleye fingerlings from ponds usually takes place during the summer. Considerable care is required to minimize stress and prevent both severe immediate and delayed losses resulting from handling stress. The following recommendations are based upon unpublished experiences and observations at hatcheries in Iowa, New York, and Pennsylvania.

1. Ponds, whether drainable or not, should be free of filamentous algae and macrophytes. Many fish tend to become trapped in vegetation during draining and seining operations.
2. Harvest operations should be conducted when water temperatures are below 20°C, if possible. If fish are concentrated in a catch basin, a flow of fresh water through the basin will greatly reduce stress.
3. Harvested fish should be immediately placed in a solution of 0.5% NaCl and an appropriate antibacterial agent should be added.
4. DO concentrations in all tanks, tubs, pails, or other transport containers should be maintained at or above 4 mg/ℓ.

Given the above conditions, walleyes can be harvested and transported for up to 6 hr with no appreciable losses according to studies conducted by the Iowa Cooperative Fishery Research Unit. The importance of extreme care in the harvest and transportation of walleyes has not been fully appreciated and cannot be overemphasized. Too often it has been assumed that as long as the fish were alive when stocked, they would survive. There is no established basis for such an assumption.

Pond production of walleye fingerlings is currently necessary since fry have not been successfully reared in large numbers under intensive hatchery conditions. However, the relatively low production that is obtained in ponds and the unpredictability of that production hinder the development of large-scale production facilities. These factors provide a major incentive for research and development of intensive culture procedures applicable to walleyes.

C. Intensive Culture Methods

The state-of-the-art with respect to intensive culture of walleyes was summarized in 1978 by Nickum,[682] and the discussion which follows draws heavily on that material, but also incorporates the results of more recent studies, both published and unpublished.

The National Task Force for Public Fish Hatchery Policy[728] identified "the inability to rear the tiny delicate larvae of species like striped bass and walleye on artificial diets" as "the most critical bottleneck in the national fish-culture program." Although progress has been made toward understanding the factors which affect feeding and the general culture requirements of larval walleye, survival rates above 5 to 6% through 1 month posthatch have not been achieved regularly on diets of formulated feeds. The conversion of pond-reared fingerlings to dry, pelleted feeds has become nearly an accepted production technique and is routinely practiced in several state hatcheries and university laboratories. The techniques involved in rearing walleye fingerlings improve yearly, but they are fundamentally similar to procedures described in the 1970s.[666,729-733] The discussion which follows is divided into procedures for various phases of culture, but it should be understood that procedures for obtaining and hatching eggs are the same in all instances as those described above.

1. Fry Culture

Walleye fry will readily feed on nonliving foods; however, survival for over a month has been uniformly low (typically less than 1%). Current studies conducted at the Iowa

Cooperative Fishery Research Unit have resulted in survival rates of 5 to 6% through 1 month, but these procedures have not been tested at production levels. A variety of factors seems to affect the feeding of intensively reared fry, but one can only speculate on the mechanisms which control unsatisfactory growth and poor survival. Rearing units, stocking densities, feeds, feeding practices, and the physical conditions which exist have all been implicated as important considerations.

a. Physical Conditions and Facilities

A variety of rearing units have been tested in intensive culture situations. They range from so-called "standard" start-troughs (about 3 × 0.4 × 0.15 m) to hatching jars. No particular unit can be recommended at this time, since no practical method for intensive culture has been developed. There are, however, a number of physical factors that have been demonstrated to affect larval walleyes.

Corazza and Nickum[734] found that walleye larvae are so strongly attracted to the sides of light-colored rearing units that they ignore all forms of live or formulated feeds. Fry in uniformly lighted units with dark or neutral-colored sides become more uniformly distributed and feed actively.

Walleye fry feed most actively when they are within the water column of their rearing unit rather than on the surface or near the bottom. They also seem to have a relatively short search radius in which they can recognize and ingest food particles. It seems important, therefore, to use rearing units designed to maintain feed particles in suspension. Various upwelling systems, usually with perforated baffle plates in the bottom, have been used to accomplish feed suspension without producing strong currents that might cause battering of the fish.

The outlets of rearing units must be screened with a fine-meshed material so larval walleye (7 to 8 mm long) do not escape. Since feed and fecal material rapidly accumulate on such screens, the outlet should be sufficiently large to prevent clogging and subsequent overflow with resultant loss of fish. Water exchange rates of at least 2/hr seem to reduce fouling and disease problems.

No controlled, replicated studies have tested the effects of temperature on feeding of walleye fry. In natural environments fry may be assumed to begin feeding at 10 to 15°C if a normal warming trend follows spawning. It seems reasonable to suggest similar temperatures for intensive walleye fry culture. Temperatures of 20°C have led to generally poor results. Temperatures above 10 to 15°C lead to increased metabolic rates in the fish and apparently cause the fry to exhaust their nutrient reserves before they can adapt to prepared feeds, thus causing massive starvation and cannibalism. Temperatures of about 20°C have been found desirable for walleye fingerling rearing, however.

Stocking density and feeding intervals have also been thought to influence the feeding of walleye under intensive culture conditions. No density guidelines can be offered as yet, but a typical density is 200,000 fry per standard trough. Some workers believe that crowding aids initial feeding, though no systematically gathered evidence supports that belief. Frequent feeding (eg. intervals of 2 to 5 min) seems to improve feed intake and survival probably due to the feed particles being fresh and within short distance from the fish.

b. Feeds and Feeding

Numerous diets and environmental conditions have been used when feeding walleye fry under intensive culture conditions. Brine shrimp nauplii have been the standard live food source. Various experimental diets, including the Oregon Moist Pellet, chicken egg yolk, trout and salmon starter diets, liver slurry, farina slurry, and mixed zooplankton have all been tested under intensive culture conditions. The most widely used formulated feed is the W-14 diet (Table 1, Chapter 6) developed by the Beulah, Wyoming Fish Technology Center (formerly Spearfish Diet Development Center, South Dakota) of the U.S. Fish and Wildlife Service. Greatest survival and growth have been obtained with brine shrimp nauplii, but no diet or rearing system can be regarded as well established.

Failures with both dry diets and live feeds have tended to take one of two forms. Immediate failures to accept feed have been followed by cannibalism, "tail-biting", and early starvation. When the feed has been accepted by a proportion of the fry, a phenomenon commonly called the "dwindles" by hatchery workers has been repeatedly observed. Fish 2- to 3-weeks old die in substantial numbers for no apparent reason and the population may fall to near zero. Malnutrition is the most probable explanation for the phenomenon. Failure of intensively cultured fry to inflate their gas bladders may also contribute to the phenomenon.[565]

Continual availability of feed appears to be important. Acceptance of dry diets has been highest when the fish are fed at 2- to 5-min intervals with automatic feeders. Kostomarova[735] suggested that "pike" fry require feed during the final stages of yolk-sac absorption to prevent mass starvation. Although it is improbable that he was working with walleye fry, it may be useful to note that a continual supply of required nutrients was critical when the fish were young.

Cuff[736] concluded that greater food availability significantly reduced cannibalism in walleye less than 20 days old, primarily because feeding fish more successfully avoided attacks by cannibals in the population. Jahncke[718] found that feeding a nutritionally balanced diet must begin within 2 to 5 days of hatching, depending upon temperature, or the fry will develop an energy/nutrition deficit from which they will not recover.

The feeding behavior of walleye fry on natural foods may provide a direction for feed development and feeding practices.[737-739] The live foods most commonly utilized typically have one axis less than 0.4 mm. Tests with various pelleted diets also indicate that walleye fry will ingest particles of 0.2 to 0.4 mm. Some evidence points to greater acceptance of orange and red particles over other colors, but this has not been confirmed. It appears important to maintain a high density of feed particles whether living or prepared feeds are used.

A diet based on the nutrient composition of unfertilized walleye eggs was readily accepted by walleye fry in recent studies conducted by the Iowa Cooperative Fishery Research Unit. The red, freeze-dried-gelatin particles were apparently adequate nutritionally, since the fish grew normally to lengths of 17 to 20 mm. However, none of the fish inflated their gas bladders and complete mortality occurred. Similar problems have been observed by other walleye culturists and by persons working with other species such as striped bass (see Chapter 9). Solution of the gas bladder inflation problem may hold a key to successful intensive culture of walleyes.

c. Pathology

Pathological problems in walleye fry culture have not been widely investigated. Problems related to the presence of myxobacteria and fungi have been reported (but not confirmed) at several hatcheries. A fungal infection involving the oral cavity and gills was prevalent in one set of rearing trials conducted in New York during 1977, and a similar problem was reported at Linesville, Pa. It is not known whether the infection was primary or secondary. The flush rate in all troughs was doubled in later trials and no further problems were observed.

Preliminary tests of the toxicity of standard therapeutic agents have been inconclusive. Walleye fry seem to be unusually sensitve to many of the chemicals commonly employed by fish culturists. If that observation is sustained, the necessity for maintaining high water quality standards will become even more important, since even chemicals approved by the U.S. Food and Drug Administration may be toxic to walleye fry at levels necessary for efficacy in other species.

d. Suggested Methods for Walleye Fry Culture

The following suggestions, while preliminary, may be helpful to culturists involved in walleye fry rearing:

1.   Cylindrical rearing units with large, finely screened outlets are preferred over standard start troughs. In either type of unit, water flow should be directed in a manner which aids in suspending food particles. Rearing units should be of a neutral color and should be uniformly lighted.
2.   At least two exchanges per hour of 10 to 15°C water should be provided. Once feeding is established the temperature may be raised to 20°C.
3.   Live feed such as brine shrimp nauplii should be provided in large quantities (1000 nauplii/fish/day seems appropriate) until swim bladder inflation occurs. Conversion to prepared feeds should be accomplished abruptly; that is, no gradual weaning period should be used. Fry reared on brine shrimp nauplii will ignore prepared feeds so long as live food is available.
4.   If dry feeds are used as the initial diet, they should be fed at 2- to 5-min intervals, 24 hr/day. A feed with nutrient content similar to that of unfertilized walleye eggs is recommended. Since feed particles must be small (0.2 to 0.4 mm) the solubility of nutrients in water must be considered and appropriate adjustments in diet formulation made. A ring to confine feed particles and oils from the feed to a portion of the rearing unit surface may improve survival.

2. Fingerling Culture

Intensive culture of pond-reared walleye fingerlings is relatively new, but methods developed in laboratory systems have been adapted for use in several production hatcheries. Methods are similar to those described by Nickum,[682] but a number of modifications have been made. Most of the modifications are based on research and development studies conducted within the last few years. Therefore, the material which follows incorporates that experience and conversations with other walleye culturists.

a. Physical Conditions and Facilities

Supplies of fingerling walleyes for intensive culture can be reliably produced in ponds using the methods described above. Highest production is obtained when the fish are harvested before reaching a length of 5 cm. Fish as small as 2.5 cm convert readily to formulated feeds if harvest and handling stress are minimized and the methods presented below are followed.

Rearing units of various sizes and shapes have been successfully used for intensive culture of walleye fingerlings. Experience to date indicates that the most important consideration involving rearing units is that the flow of water should be directed in a manner which will keep feed particles in suspension. Upwelling systems or water flows introduced perpendicular to the long axis of the culture unit have been successfully employed. Water exchange rates should be adjusted to reflect the oxygen demand of the fish and the feeding rate. Heavy stocking densities and high feeding rates may require three to five exchanges each hour; however, a single exchange hourly is satisfactory for lower stocking rates. At present, 25 kg/m³ is considered maximum stocking density.

Walleye fingerlings should be harvested and transported at temperatures below 20°C, though they will survive and grow in intensive systems up to 30°C. Temperatures of 20 to 25°C seem optimal for good growth and low disease susceptibility. Temperatures below 18°C reduce growth rates and growth is negligible below 10°C. Dim light seems to enhance feeding and the fish are less excitable if they are kept in covered units where disturbance by passersby is reduced. Photoperiods of 16 to 24 hr light daily seem appropriate.

Waters of widely varying quality have been used for walleye culture. Water clarity may be of some importance, but other factors do not seem to influence walleye growth and survival, so long as stressful limits are not exceeded. Water of low turbidity generally leads to fewer losses attributable to cannibalism; however, it is not known whether that result is related to changes in walleye behavior or to the ability of the culturist to more closely observe the behavior of cannabilistic fish. Clear, clean water is also recommended because of reduced disease incidence.

b. Feeds and Feeding

It is now considered desirable to start pond-reared fingerlings which are moved to intensive culture systems directly on dry pellets with no weaning period. Starter size granules of such diets as W-14 and W-16 (Table 1, Chapter 6) supplied through automatic feeders at 2- to 10-min intervals, 16 to 24 hr/day, have been readily accepted by walleye fingerlings. Within a week, 60 to 80% (survival as high as 99% through 1 month has been obtained) of 3- to 5-cm fish can be expected to accept such prepared feeds. If nonfeeding fish are removed and isolated in separate tanks, many of them will learn to accept feed. Particle size should be increased as the fish grow.

No published information on feeding rates and feed conversion ratios is available. Experimental programs in which walleye fingerlings have been fed to satiation have typically resulted in feeding rates of 5 to 10% of body weight daily. Feed conversions as low as 1.5 have been obtained in such studies, but many variables are known to influence the values obtained.

The W-series of diets developed at U.S. Fish and Wildlife Service laboratories contain protein levels in excess of 55% and are, therefore, quite expensive. Preliminary results from studies by the Iowa Cooperative Fishery Research Unit indicate that protein percentage can be reduced to 45% with no sacrifice of growth. Substantial modifications in diet specifications can be expected now that it is possible to reliably produce pellet-fed fingerlings for such studies.

Variations in feed color and texture, as well as flavor enhancers have been used to increase the acceptance of dry feeds developed for walleye fingerlings; however, most studies have lacked adequate replication and controls. It is the opinion of most workers that soft feeds are more acceptable than hard pellets. No consistent effects attributable to flavor enhancers have been reported.

c. Pathology

Myxobacterial infections, particularly those attributable to *Flexibacter columnaris*, have been the primary pathological problems associated with walleye fingerling culture. *Ichthyophthirius multifiliis, Trichodina* sp., *Scyphidia* sp., bacterial gill disease, furunculosis, fin rot, and fungal infections have also been reported.[740] At least 68 additional parasites and diseases of walleyes in natural populations have been identified,[741] but the majority do not appear to be serious threats in intensive culture systems.

Various treatments have been employed to combat pathological problems in walleye culture. Nagel[732] controlled outbreaks of *F. columnaris* with 10-sec dip treatments in 500 mg/ℓ copper sulfate and 1-min baths in copper sulfate at 30 mg/ℓ. Hyamine 3500 at 2 mg/ℓ for 45 min and Diquat at 16 mg/ℓ have also been reportedly successful against *F. columnaris*.[740] Malachite green, formalin, Acriflavine, potassium permanganate, and Furanace have been used with variable success. Hnath[740] reported that Roccal at 2 mg/ℓ for 1 hr effectively controlled bacterial gill disease. He also indicated that Acriflavine at 5 mg/ℓ for 1 hr or Hyamine 3500 at 2 mg/ℓ for 1 hr controlled fin rot and that Terramycin in feed controlled the symptoms of furunculosis. Formalin treatments for *Ichthyophthirius* were not successful.[740] Nagel[731] controlled bacterial gill disease with Roccal at 2 mg/ℓ and fungus disease with formalin at 1:6000.

Disease problems with fingerling walleyes have apparently been reduced by minimizing handling when water temperatures exceed 20°C, by strict sanitation, and by maintaining high water quality and rapid flush rates. Dietary insufficiencies may contribute to disease problems; however systematic studies of pathology association with diet have yet to be conducted.

d. Suggested Method for Fingerling Production

The methods suggested below should lead to acceptable survival and growth of walleye fingerlings under intensive culture conditions. However, as with any new technology, modifications of the method will be required to reflect advances in research and practical experience.

1. Walleye fingerlings should be harvested from ponds when they reach 2.5 to 3.5 cm, a range in which starvation and cannibalism losses are low. Harvest should be undertaken when water temperature is below 20°C, and stress should be minimized through careful handling and the use of 0.5% NaCl and a bacteriostatic agent in all transportation units.
2. Rearing tanks designed to maintain feed particles in suspension should be used. Water flows of one to two exchanges hourly of 20°C water are recommended. Covered tanks or troughs and dim lights for at least 16 hr daily are also recommended.
3. Feed with formulations similar to W-14 or W-16 should be provided at 2- to 10-min intervals at least 16 hr/day. A feeding rate of 10% of body weight daily may be needed during initial feeding, but 3 to 6% daily should be adequate once the fish are actively feeding.
4. The fish should be carefully observed at frequent intervals. Cannibals should be removed and the fish graded at regular intervals if differential growth is observed.

3. Post-Fingerling Culture

Walleyes will continue to grow year-round if water temperature is maintained above 20°C. Growth will continue at temperatures down to 10°C, but becomes very slow below 15°C. Length increases of 5 cm monthly and doublings in weight at 2-week intervals have been obtained, but should not be expected on a sustained basis; 2-year-old fish with well-developed gonads have been produced under laboratory conditions.

The methods used to rear walleyes past the fingerling stage are essentially the same as those used for fingerling production. Pellets of 8-mm diameter will be accepted by walleyes of all sizes beyond 20 cm (and by some smaller individuals). All types of rearing units, including floating net pens (cages) have been used on an experimental basis. Specific optimal rearing conditions and diets have not been developed, but the diet, facilities, and conditions used for fingerling production are generally adequate.

IV. SUMMARY

Intensive rearing of walleyes is limited, at present, to fish of 2.5 cm length or greater. Pond production systems must be used to rear substantial numbers to that size; however, pond production becomes unreliable when used to produce fish greater than 5 cm long.

Chapter 9

STRIPED BASS AND STRIPED BASS HYBRIDS

Jerome Howard Kerby

TABLE OF CONTENTS

I. INTRODUCTION

The striped bass (*Morone saxatilis*) is a desirable sport and commercial species which is anadromous, euryhaline, and most commonly found in marine and estuarine waters (Figure 1). However, south of the Roanoke River, North Carolina, and along the Gulf of Mexico coast, it is basically a riverine species.[742-744] Along the Pacific coast, the Sacramento-San Joaquin population normally inhabits the river deltas or San Francisco Bay, but sometimes makes coastal migrations in patterns that appear to correlate with warmer waters.[745]

Landlocked populations of striped bass have been introduced to reservoirs across the U.S.[746,747] The range of the species, which extends from the St. Lawrence River, Canada, to the St. John's River, Florida on the Atlantic coast; from British Columbia to south of the Mexican border on the Pacific coast; and along the Gulf of Mexico coast, indicates that the species has a wide tolerance for a variety of environmental variables.[745,748,749]

Although hybridization in fishes has been widely studied, only in recent years has artificial hybridization been recognized as a tool for the potential improvement of fish stocks and for management purposes. Hybrids of female striped bass and male white bass (*M. chrysops*), here termed SB X WB, were first produced by Robert E. Stevens in 1965.[750] Several other crosses have also been made, but none has gained the acceptance of the original.[750-754] Superior growth rate (particularly during the first 2 years of life), greater disease resistance, improved survival, and general hardiness have been demonstrated by *Morone* hybrids.[750,753-758] Since 1965, the SB X WB hybrid, and to some extent its reciprocal, have been artificially propagated and widely stocked in freshwater impoundments for control of shad (*Dorosoma* spp.), and as food and sport fish. In 1981, about 40 million *Morone* fingerlings were produced in 17 state and federal hatcheries for stocking inland waters. Over 450 reservoirs (2.3 million ha) have been stocked.[747,759]

During the past 2 decades breakthroughs in spawning and rearing techniques, coupled with stocking of striped bass in large reservoirs have led to new recreational fishing and management opportunities. Recent research has also suggested that the potential exists for commercial aquaculture of striped bass. This chapter reviews the history of striped bass and striped bass hybrid culture and discusses some of the associated problems and potentials.

II. EARLY HISTORY

Captain John Smith noted that there were, "such multitudes [of striped bass] that I have seene stopped in the river close adjoining to my house with a sande at one tide as many as will loade a ship of 1000 tonnes."[760] However, by the 1880s, population declines were already causing concern among fisheries workers.[761] Consequently, the first striped bass hatchery was established in 1884 at Weldon, N.C., where ripe fish could be obtained from natural spawning areas.[762] In the first year of operation some 298,000 larvae were hatched and 280,500 were released into the Roanoke River. Ripe adults were supplied to the hatchery by fishermen.[761] The hatchery has operated almost continuously during the intervening years.[763]

In 1941, Pinopolis dam was closed to form Santee-Cooper reservoir in South Carolina. By 1950, appreciable numbers of striped bass had appeared in sport catches, and it soon became clear that saltwater was not a physiological requirement for reproduction.

The successful population in Santee-Cooper reservoir fed primarily on clupeid species such as gizzard shad, *D. cepedianum*.[764-766] That finding prompted fishery man-

FIGURE 1. Comparison of striped bass X white bass hybrid (above) with striped bass. The hybrid was 395 mm in fork length (FL) and 1058 g. The striped bass was 398 mm FL and 824 g. (From Kerby, J. H., Burrell, V. G., Jr., and Richards, C. E., *Trans. Am. Fish. Soc.*, 100, 787, 1971. With permission.)

agers to introduce striped bass as a biological control species in reservoirs where shad populations were causing problems. Although the stocked fish usually thrived, they did not reproduce,[746,767-769] and it was concluded that most reservoir systems did not provide the proper spawning environment — sufficient upstream river length and currents strong enough to keep the semibuoyant eggs in suspension until they hatched.[770]

The lack of spawning success in most reservoirs led to the need for additional hatcheries. The potential for success of those hatcheries was greatly improved following the development and refinement of a satisfactory technique for hormone-induced spawning,[770-772] which rendered the capture of naturally ripe females unnecessary.

III. LARVAL PRODUCTION

Two principal methods are used to produce striped bass larvae. In both, broodfish are collected and transported to holding facilities. In one method the females are injected intramuscularly with 275—300 IU of human chorionic gonadotropin (HCG) per kilogram of body weight. Males may also be injected with 110 to 165 IU/kg of HCG to increase semen volume. Between 20 and 28 hr after injection, a small egg sample is taken with a 3-mm outside diameter glass catheter inserted through the urogenital opening into the ovary (Figure 2). Approximate time to ovulation is determined by microscopic examination of the eggs (Figures 3 to 6). Actual time of ovulation is veri-

FIGURE 2.   Female striped bass catheterized with a 3-mm diameter glass tube to obtain an egg sample for predicting time of ovulation. (From Jack D. Bayless, South Carolina Wildlife and Marine Resources Department.)

fied when freely flowing eggs are produced after pressure is exerted on the abdomen. Accurate prediction is important because the eggs separate from the ovarian tissue and from the parental oxygen supply at ovulation. Anoxia results within about an hour if spawning does not take place.[768,771-772]

Female broodfish are either sacrificed or anesthetized by spraying the gills with a 0.1% solution of quinaldine. The eggs are then manually stripped into a spawning pan (Figure 7). Semen is stripped from males into the pan, water is added, and the contents of the pan are mixed. Following fertilization, the eggs are placed in modified McDonald hatching jars (Figure 8) at the rate of about 100,000 eggs per jar. A continuous flow of oxygenated water is introduced at a rate sufficient to keep the eggs in motion. After 40 to 48 hr, depending upon temperature, the larvae hatch and swim over the lip of the jar into an aquarium. Larvae are normally held in aquaria for 2 to 5 days before being shipped or stocked.[757,771,772]

The second larval production method involves tank spawning[773] (Figure 9). Both

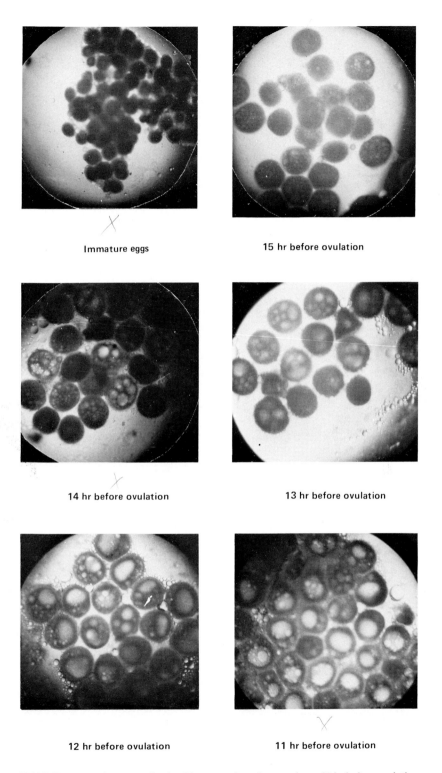

Immature eggs

15 hr before ovulation

14 hr before ovulation

13 hr before ovulation

12 hr before ovulation

11 hr before ovulation

FIGURE 3. Development of striped bass eggs from immaturity to 11 hr before ovulation. (Courtesy Jack D. Bayless, South Carolina Wildlife and Marine Resources Department.)

10 hr before ovulation
Polarization complete

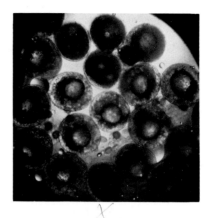

9 hr before ovulation
Nucleus clearing

8 hr before ovulation

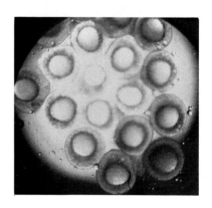

7 hr before ovulation

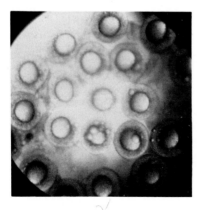

6 hr before ovulation

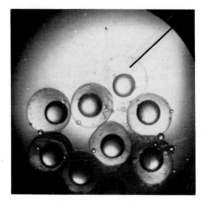

5 hr before ovulation

FIGURE 4.   Development of striped eggs from 10 to 5 hr before ovulation (magnification × 20). (Courtesy Jack D. Bayless, South Carolina Wildlife and Marine Resources Department.)

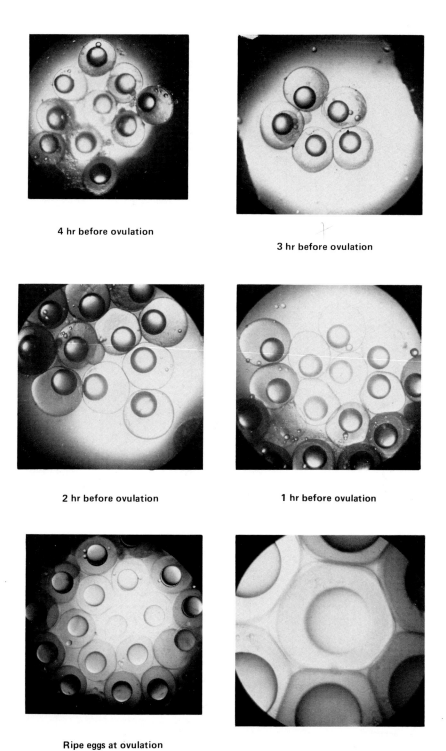

4 hr before ovulation

3 hr before ovulation

2 hr before ovulation

1 hr before ovulation

Ripe eggs at ovulation

Ripe eggs at ovulation (50X)

FIGURE 5. Development of striped bass eggs from 4 hr before ovulation to ripeness (magnification × 20). (Courtesy Jack D. Bayless, South Carolina Wildlife and Marine Resources Department.)

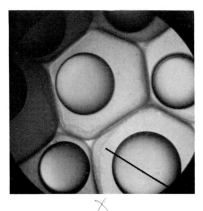

Overripe eggs 1 hr (50X)
Note breakdown at inner
surface of chorion

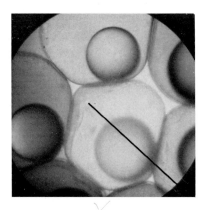

Overripe eggs 1½ hr (50X)
Breakdown at inner surface
of chorion persists

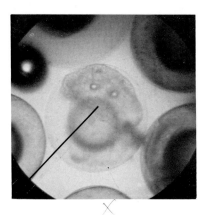

Overripe eggs 2 hr (50X)
Note deterioration confined
to one-half of egg

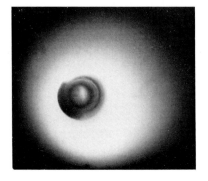

Overripe egg 16 hr (20X)
(Dark areas appear white
under microscope)

FIGURE 6.   Development of striped bass eggs that have passed the ovulation stage and have become overripe. (Courtesy Jack D. Bayless, South Carolina Wildlife and Marine Resources Department.)

males and females are injected with HCG at the rates outlined above. Normally, two females and four males are placed in a circular tank 1.2 to 2.4 m in diameter and about 1.2 m deep. Water is supplied through two or more 13-mm inside diameter tubes under slight pressure at the rate of 30 to 38 ℓ/min, creating a circular velocity of 10 to 15 cm/ sec at the tank perimeter. A 0.1-m diameter center standpipe encircled by a fine-mesh screen (0.45 m in diameter) controls water depth and prevents the loss of eggs and larvae. A bubble-curtain around the base of the screen keeps eggs and larvae from becoming entrapped. The fish normally spawn from 36 to 62 hr after injection, depending upon water temperature. After spawning, the broodfish are removed from the tanks and released. The larvae are subsequently collected with fine-meshed scoops or a siphon attached to a large funnel.[773]

The jar method is the more labor intensive of the two, requires more expertise in predicting time of ovulation, and is the more expensive. At least three persons are required on a 24 hr a day basis for efficient operation of a jar hatching facility. There

FIGURE 7. Ripe eggs from a female striped bass are stripped into a spawning pan.

is also more stress and even mortality of broodfish when strip spawning is used as compared with tank spawning.

Because jar culture requires less space than tank culture, production per unit area is higher in the former. Larval numbers are also easier to predict when the jar method is employed, and the culturist has greater control over the developing eggs and larvae than when they are in tanks. A major disadvantage of the tank culture method is that hybrids cannot be produced because female striped bass, even after they have ovulated, do not release their eggs in the presence of male white bass.[773]

IV. FINGERLING PRODUCTION

Attempts in the 1960s to establish striped bass populations in reservoirs by stocking larvae were largely unsuccessful. More than 25 reservoirs in 9 states were stocked with larvae, many in two or more successive years. In only two reservoirs were fisheries eventually established. One of those populations was composed of SB X WB hybrids.[746,774] Thus, it was concluded that successful fisheries are dependent upon the stocking of fingerlings.[769] However, rearing the larvae proved to be more difficult than was originally envisioned because they are highly sensitive to a variety of factors. At certain stages, they "shock" easily and the reasons for their deaths are not readily discernible.

Ponds are presently most widely used for fingerling production, but more intensive tank and raceway culture is also being successfully employed. Both fresh and brackish water have been utilized effectively.

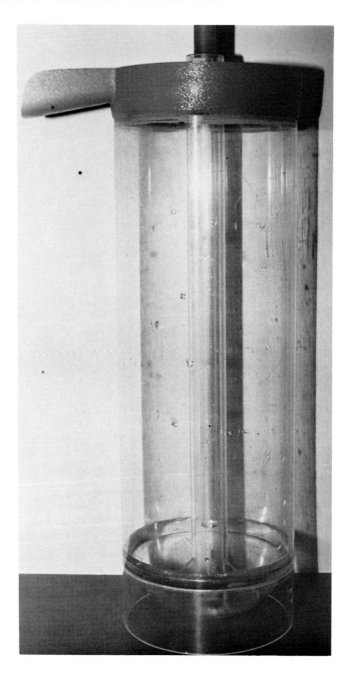

FIGURE 8.   Modified McDonald hatching jar with a "tube-within-
a-tube" arrangement to prevent air bubbles from floating eggs out of
the jar.

A. Pond Culture

Success in rearing striped bass larvae to fingerlings first occurred in 1964 at the
Edenton, N.C. National Fish Hatchery, where about 30,000 fingerlings were produced.[775] Subsequently, personnel at the Edenton hatchery explored and developed
new culture methods and annually increased production.[776-778] Data collected by various hatchery personnel were synthesized into comprehensive guidelines concerning
striped bass culture and published in 1976.[757]

FIGURE 9.   Circular tank arrangement for allowing striped bass to spawn naturally. (Courtesy R. David Bishop, Tennessee Wildlife Resources Agency.)

Larval survival to the fingerling stage was normally less than 20% and often below 10% during the early years of culture.[769,774,779-782] Even with recent advancements in culture methodology, most hatchery managers are presently satisfied with 30% survival.[783]

Most culture currently takes place in freshwater earthen ponds, though brackish water is used in some areas. Ponds range in size from less than 0.04 to more than 2 ha, but most are in the 0.1- to 0.4-ha range and are usually 0.9 to 2.5 m deep.

Ponds are normally dried and disked during winter to promote the degradation of organic matter. Rye grass or other cover is sometimes planted to reduce erosion and serve as a source of organic fertilizer,[757,784] but should be used with caution. Use of chemicals such as lime or herbicides may be advisable, depending on local circumstances. Pond filling is normally not recommended until shortly before stocking to reduce the effects of predacious insects and to promote the availability of small zooplanktonic organisms. Copepods and cladocerans are important to striped bass until they reach 80 to 120 mm.[769,780,785-788]

Organic fertilizers are generally preferred over inorganic types because the slower decay of organic materials seems to provide more sustained zooplankton production. Inorganic fertilizers, especially when used alone, often produce dense phytoplankton blooms that may result in high pH and can cause oxygen depletion, a dominant blue-green algae bloom, or both.[757,784]

The most commonly used organic fertilizers are hay; alfalfa, bermuda, or milo pellets (or meals); and soybean or cottonseed meals. Peanut, alfalfa, bermuda, and lespedeza hays are often used. Hay is applied at higher rates than pellets and meals; soybean and cottonseed meals are applied at the lowest rates. Many culturists either soak or grind hay in a hammer mill prior to application to hasten decomposition. Subsequent applications may be required to maintain plankton blooms.[757,789]

Studies which compared ponds fertilized with ground bermuda hay with those fertilized with poultry waste combined with liquid inorganic fertilizer demonstrated that the latter treatment supported a larger and more diverse forage base with less aquatic vegetation than the former.[530,726] Furthermore, striped bass fry survival was increased in the poultry waste-liquid fertilizer treatments. Recommendations included combining organic and liquid inorganic fertilizers, combined with pond aeration, to optimize zoo-plankton production. In addition, applications of lime to enhance the potential biological activity of the inorganic fertilizer was recommended in cases where alkalinity is low and the water is soft. Selective inoculation of desirable crustacean zooplanktonic species and unicellular green algae was also recommended, especially in instances where pond water is obtained from wells or springs. Timing and quantities of fertilization and inoculation of plankton should be shifted to maximize crustacean zooplankton and their foods during the 2-week periods before and after stocking of the fish.[530,726]

Striped bass larvae have normally been stocked into ponds at 5 to 10 days of age, following swimup and an initial period of feeding on brine shrimp (*Artemia salina*) nauplii. Vessels for holding larvae until release into ponds include aquaria, raceways, and cages placed in ponds. There may be enhancement of pond production subsequent to holding larvae in cages as compared with holding them in aquaria.[790] Recently, the practice of holding larvae for any extended period of time has been questioned and some agencies are presently stocking fish at an age of 5 days or less.[791]

While high larval stocking rates yield more fingerlings per hectare up to a point,[780,792,793] all systems have maximum carrying capacities and excessively large numbers of fish may lead to correspondingly smaller average sizes.[793] The current recommendation for stocking is 247,000 larvae per hectare.[757]

Fingerlings are normally harvested 30 to 50 days after stocking at a size of 25 to 50 mm and weight of 0.3 to 1.5 g (Phase I fingerlings). They may be stocked into reservoirs or restocked into culture ponds until fall or winter (Phase II fingerlings). The latter fish are fed prepared diets, usually trout or salmon pellets. Phase II fingerlings may reach 170 mm by later winter.[794]

B. Intensive Culture

Culture of striped bass to the fingerling stage in aquaria, troughs, tanks, or raceways has nearly as long a history as pond culture.[763,775] Intensive culture has some advantages as well as problems. Such systems are space efficient, can be precisely controlled, allow for easy observation of the fish, and for close monitoring of water quality. In addition, disease treatment is facilitated. On the negative side, intensive systems require a constant and adequate supply of high quality replacement water, the quality of which must be maintained. Most closed systems are relatively complex, and the threat of mechanical or electrical failures requires extensive alarm and backup systems. Cleaning of tanks and filters can be a major problem and may be labor intensive. Because of the high densities of fish being maintained in such systems, disease and cannibalism can become significant problems, and provision of sufficient quantities of appropriate foods may be difficult prior to the Phase I stage.

Survival rates in early intensive culture systems were often less than 1% and seldom exceeded 10%.[672,763,775,795-797] Technological advances have now led to systems in which more acceptable survival percentages are routinely achieved.[672,796,798,799] Average survival has increased from less than 10% in 1974 to 46% in 1980 in one system.[672]

Intensive systems for the production of Phase II fingerlings have been more successful than for production of earlier stage fish, primarily because fingerlings 30 to 40 mm long can be readily trained to accept prepared feed. Survivals of 81 to 99% have been reported from systems producing Phase II fingerlings.[797,800]

C. Food Requirements

Fingerling production appears to be directly proportional to the type and abundance of zooplankton available, provided the striped bass larvae are healthy and water quality is maintained.[789] The preferred foods of striped bass larvae and early juveniles in freshwater ponds are copepods, cladocerans, and insect larvae. Fish size has a significant effect on food selectivity. Early postlarvae select the early instars of cladocerans and copepods; larger organisms are avoided.[769,785,787,794] In general, copepods constitute the most important food of striped bass less than 30 mm long.[776,786,788]

Cladocerans become increasingly important as the fish grow. Highest utilization occurs when the fish are in the 20- to 80-mm length range. Insects are important to fish of all sizes, but become dominant foods for fish over 80 mm.[776,780,785·787,801] In wild populations, forage fish become important foods by the time striped bass reach 80 to 90 mm.[801,802]

Food preferences of striped bass reared in brackish ponds have not been described though salinity seems to affect food habits in natural estuarine situations.[803,804] As salinity increases, the dominant copepod species in the diet change from *Cyclops* and *Diaptomus* to *Eurytemora, Acartia,* and *Pseudodiaptomus.*[803] Amphipods, mysids, isopods, and polychaetes are also consumed in saline waters.[803,804]

Intensive culture presents problems since insufficient zooplankton of the preferred types can be produced in a secondary culture system and early postlarval striped bass will not readily accept prepared diets.[805·807] Though expensive, brine shrimp nauplii are readily accepted by striped bass; thus, they are widely used in intensive culture. *A. salina* nauplii seem to provide adequate nutrition, though there have been reports of nutritional problems after metamorphosis.[776,784]

A clearly defined critical period for striped bass survival does not occur if sufficient food is made available.[808·810] Unfed larvae decrease in length and lose weight, but most survivors are able to capture and consume food even when they are near death.[810] Survival of over 70% has been reported when food was offered on or about 18 days after egg fertilization.[809] Brine shrimp nauplii density is a key in promoting good survival and growth of striped bass larvae.[672,809,811] Initial nauplii concentrations of 50 to 60/mℓ, followed by increased food density to 100 to 120/mℓ when the fish reach 15 days of age have been successfully employed.[672]

Feeding should begin as soon as the digestive tract is developed, usually 4 or 5 days after hatching, depending upon temperature. Feed should be provided at least every 3 hr, and continuous feeding is recommended.[757,812]

Continued survival beyond metamorphosis requires that the fish be trained to accept prepared rations. Prepared food with about the same particle size as brine shrimp nauplii may be provided when the fish are between 14 and 21 days of age.[757] Both prepared feed and brine shrimp should be fed until the 28th day of age, after which brine shrimp can be phased out. Alternatively, 12-day-old larvae can be trained on flaked particles smaller than brine shrimp nauplii until the fish reach 17 days of age. Thereafter, a high protein, pulverized salmon starter can be employed exclusively. Frequent feeding — 12 to 16 times daily — has been recommended, during which feeding rates of 25 to 50% of body weight daily may be employed. Feeding rates can be gradually reduced to 5% of body weight daily when the fish are 40 to 45 days old.[672]

A wide variety of commercial feeds are available, including various brands of high protein trout and salmon rations. A combination of dry trout feed and whole processed fish make a good diet when mixed and reground.[757] Various types of ground fish, shrimp, and liver can also be fed, but water quality problems may subsequently occur, particularly in intensive culture systems.

Nutritional requirements have not been well defined for striped bass, but salmon rations containing at least 38% protein are generally considered adequate.[757,813] Two rigorous studies of striped bass nutrition examined dietary protein level and the interactions between dietary protein and lipid with respect to effects on growth, food conversion, and protein utilization.[814,815] When formulated diets containing 34, 44, and 55% crude protein were fed to juvenile striped bass, the diet with the highest protein percentage led to highest average weight gain and best food conversion. Some protein sparing was observed when lipids were added at 7, 12, and 17% of the diet.

D. Predation

If care to exclude other fish species is exercised by filtering surface water prior to use, there are three primary sources of predation on cultured striped bass: predaceous insects, cannibalism, and fish-eating birds. Aquatic insects, particularly back swimmers (Notonectidae) and phantom midge larvae (*Chaoberus* spp.) are the principal predators on larvae and early juveniles reared in ponds, whereas cannibalism can become significant in intensive culture systems. In addition, predation on small striped bass larvae by a free-living copepod (*Cyclops bicuspidatus thomasi*) has been noted in brackish natural waters[816] and could become a problem in culture. Birds such as kingfishers, gulls, herons, and ospreys may become important predators in ponds, outdoor tanks, and raceways after the fish become juveniles and subadults.

Newly hatched striped bass larvae are particularly vulnerable to predation because they cannot swim continuously.[794] Phantom midge larvae have been observed to kill striped bass larvae in the laboratory and could post a serious problem in ponds.[763] Several methods of controlling predaceous insects are available. One accepted method is to drain ponds before stocking and treat them with potassium permanganate.[757,779,784] Filling ponds immediately prior to stocking will allow the fish to grow significantly before insects become established, and application of diesel fuel to ponds leaves a surface film that leads to the suffocation of air-breathing insects. Chemicals such as Baytex and Dylox have also been used under certain circumstances.[757]

While cannibalism may occur in culture ponds,[772,788] it has not been confirmed as a serious problem. However, cannibalism can clearly become the cause for significant levels of mortality in intensive culture systems.[672,757,795-797,817,818] High larval density, perhaps coupled with inadequate or insufficient food, seems to promote cannibalism.

V. FOODFISH PRODUCTION

During 1981 to 1982, wholesale prices for striped bass in Atlantic coast fish markets ranged from $2.54 to $7.72/kg depending upon size, location, and time of year.[819] Recent declines in natural populations have led to sustained high prices and served as an inducement to groups interested in commercial culture.

The Edenton National Fish Hatchery is the only facility that has routinely reared striped bass beyond the advanced fingerling stage. Several different races of striped bass have been reared to adulthood in ponds and maintained as brood stock. Other organizations, both governmental and private, are currently exploring methods for rearing striped bass and striped bass hybrids to commercial size. Although there have been varying degrees of success, economic feasibility has yet to be demonstrated, but various aspects of the opportunities which exist for marketing the fish have been described.[820]

Marine Protein Corporation was perhaps the first organization to attempt commercial striped bass culture. Phase I fingerlings were produced in ponds and reared to advanced fingerling and subadult sizes in raceways and silos. Over 4500 kg of advanced fingerlings (254 to 279 mm) and over 9000 kg of food fish (356 to 457 mm) were produced in 6 silos during 1973 and 1974. However increased costs associated with the energy crisis of that period made the venture uneconomical.[821]

Striped bass and SB X WB hybrids can be grown to commercial size in production quantities. One option for culture is to rear the fish in cages. A study conducted in an Albama estuary[800] demonstrated survival over a 60-day rearing period in excess of 98% and standing crops exceeding 89 kg/m³. Fish in another cage study conducted in a saline New York lagoon suffered 80% mortality because of exposure to low winter temperatures.[813] In neither study was the maximum carrying capacity of the cages reached.

Net pens were employed for the culture of both striped bass and hybrids in the Stono River estuary, South Carolina beginning in 1978.[822] Hybrid survival was relatively low (16%) the first year primarily because the fish did not convert well to prepared feed. Growth of those which did survive was excellent; the fish averaged 523 and 956 g, respectively, at approximately 15 and 24 months of age. Striped bass which had been trained to accept prepared feed demonstrated 41% survival during the first year. Most of the mortality was attributed to vibriosis. In a subsequent experiment, hybrid survival at 14 months of age was 88% and standing crop was 15.8 kg/m³, whereas striped bass survival was 56%. However, growth rates were lower, perhaps due to suboptimal feeding rates.

Pond production potential for striped bass has been examined in Illinois,[823] where 10-g fingerlings were stocked in August and harvested after 441 days. Survival was 91.8%, mean weight 406 g, and standing crop 1009 kg/ha. Average food conversion ratio was 2.7. The fish went off feed when the water temperature dropped to 7°C in November and did not begin feeding in the spring until the water reached 16°C.

Culture of SB X WB hybrids to commercial size in three systems has been accomplished by North Carolina State University researchers.[824-827] Hybrids have been reared in 0.1-ha freshwater earthen ponds, 38-m³ flow-through circular pools, and 0.5-m³ floating cages placed in an estuary. Results of studies in the three systems have been variable, but generally good survival and growth occurred so long as oxygen depletions and other problems did not occur. Production as high as 5500 kg/ha has been obtained in the pond experiments.

Results of various studies on both striped bass and striped bass hybrids indicate that the fish can be grown to market size within 15 to 18 months. Ongoing research at North Carolina State University indicates that even more rapid growth is possible.

VI. WATER QUALITY

Water quality requirements necessary for striped bass culture have not been clearly defined, though the effects of specific variables have been examined and the generally favorable ranges in water quality, plus lethal limits for several variables have been reviewed with respect to larvae.[794] Striped bass appear to have a wide tolerance for several variables, and in general, the fish become tolerant to a wider range of water quality with increasing age.

A. Temperature

Temperature is one of the most important factors affecting striped bass survival and growth at all stages of development. Optimum spawning range for striped bass in the Roanoke River, North Carolina, is 16.7 to 19.4°C, though spawning occurs at temperatures as low as 12.8°C.[828] Mature ovaries subjected to temperatures of 21 to 24°C were reported to change from pale green to orange and then become opaque as the eggs die.[743] In ripe males subjected to the same temperatures, live sperm continues to be present, but motility is greatly reduced.

Striped bass eggs and larvae reportedly survive temperatures from 12.8 to 23.9°C. Outside that range, mortality increases dramatically.[828-832] However, percentage hatch begins to decline above 18.9°C,[772] and larvae hatched above 21°C may experience mortality within 70 hr.[833] Total larval mortality has also been reported at exposure temperatures above 26°C and below 10°C.[830]

Larvae and fingerlings exposed to temperature extremes survive better if they are acclimated. In general, larvae acclimate to lower temperatures more quickly than fingerlings, but acclimation to higher temperatures appears to occur at the same rate and is more rapid than acclimation to lower temperatures for both life stages.[834] Fingerlings reared in saline water tolerated temperatures as low as 6°C without acute detrimental effects.[813]

Recommended water temperatures for larvae until 9 days of age are between 4.4 and 21.1°C and pond temperatures between 18 and 32°C are considered suitable for rearing, though growth is retarded at low temperatures.[757] Growth of striped bass is reportedly most rapid and maintenance ration lowest for 100- to 286-g fingerlings when the temperature is near 24°C, and no growth occurs at 33.5°C.[835] Finally, hybrid striped bass appear to be more thermally adaptable than striped bass, and optimum temperatures for growth of the hybrids seem to be higher, reportedly as high as 31°C.[836]

B. Dissolved Oxygen

Dissolved oxygen (DO) is a major factor in striped bass culture. Mean egg survival reportedly decreases when DO falls to 5 mg/ℓ at 18.4°C, and is 50% of control survival when DO is 4 mg/ℓ at 22.2°C.[837] Incubation time increases at low DO concentrations and subsequent larval survival is reduced in direct proportion to the length of time eggs are exposed to low DO. A number of abnormalities have been reported in larvae from eggs incubated at 3 and 4 mg/ℓ DO.[837,838] Developmental abnormalities in embryos maintained at DO concentrations below 3 mg/ℓ include truncation (club tail) and scoliosis. Total egg mortality has been reported at DO concentrations near 1.5 mg/ℓ.[838]

Juvenile striped bass reportedly avoid DO concentrations of 4.9 mg/ℓ and lower.[839] Behavioral modifications of juveniles exposed to low DO include restlessness, constant swimming in an increasingly violent manner, and then spasmodic bursts to the water surface. Subsequent activities include resting on the bottom, loss of equilibrium, and finally death.[840] The first indications of restlessness occur at DO concentrations of 1.3 to 3.7 mg/ℓ. Death occurs from 0.5 to 1.4 mg/ℓ DO.

Striped bass fingerlings in ponds have been reported to survive DO levels as low as 1.4 mg/ℓ.[841] DO concentrations below 4.0 to 4.5 mg/ℓ, though perhaps not fatal, lead to reduced food consumption, increased energy expenditure for respiration, and reduced growth.[842]

C. Salinity

Survival of striped bass eggs and larvae appears to be enhanced by low salinity (about 1.7 ppt) water, and moderate salinities (8.3 to 8.6 ppt) do not appear to be detrimental.[829] Egg survival after water hardening in freshwater appears to be higher when the eggs are subsequently transferred to salinities of no more than 10 ppt as compared with transfer to higher salinities.[837]

Low salinity also appears beneficial to the survival of larval striped bass. In one study, larval growth is most rapid at 20 ppt, but survival is highest at 13 ppt. Tolerance to increased salinity improves with age, and striped bass larvae can be reared in full strength seawater following metamorphosis without apparent stress.[843] Mortality may actually be reduced when larvae are placed in tanks of static brackish water 12 hr after hatching.[844]

In experiments designed to examine interactions between salinity and temperature, salinity (up to 10 ppt) did not affect hatch percentage, but had a positive influence on survival.[832] Hatching appears to be related to temperature-salinity interactions,[832] though larval survival beyond 5 days of age may be independent of such interactions.[831] An apparent salinity-temperature interaction has been reported for growth rates of SB

X WB hybrids in brackish water.[826] In general, higher temperatures were required to produce similar growth rates at increasing salinities.

Both striped bass and their hybrids are tolerant of high salinity once the juvenile stage is reached. Striped bass X white perch hybrids acclimated to 25 ppt have been reared for 173 days without mortality.[758] In addition, SB X WB have been cultured in salinities up to 28 ppt.[822] Juvenile SB X WB and WB X SB hybrids can tolerate direct transfer to 36 ppt and subsequent return to freshwater.[845]

Striped bass are more resistant to ammonia toxicity in brackish (about 11 ppt) water at 23°C than in either fresh or undiluted seawater.[846] At 15°C, resistance is the same in fresh and brackish water, but is reduced in seawater.

Low salinities have been used in attempts to improve survival and reduce stress during harvest, handling, and transportation. Mean survival of striped bass larvae to the fingerling stage is reportedly higher in ponds brought to 1 ppt with sodium chloride than in untreated ponds.[781] Sodium chloride added to hatchery ponds at harvest may facilitate handling.[782] A 10 ppt reconstituted seawater solution has been recommended for reduction of osmotic shock caused by handling stress during transportation of striped bass fingerlings.[757] The use of salt in recirculating systems is not generally recommended because of corrosion problems and reduction in the efficiency of biofilters.[672]

D. Alkalinity, pH, and Total Hardness

Striped bass pH tolerance has not been well defined, though it is known that larvae are sensitive to pH changes.[830] The 24-hr LC_{50} for striped bass fingerlings is a pH of 5.3.[763] Juvenile striped bass prefer water which is slightly alkaline and avoid water with pH below 6.6. Aversion to acid conditions is so great that the fish will select areas with low DO levels (3.9 mg/ℓ) to remain at a favorable pH.[839] The upper lethal pH limit has been placed at 10.0,[794] but there is a report of high striped bass survival in ponds with that pH level.[776] Values of 7.5 to 8.5 seem to be optimum for striped bass culture; pH should not be outside the range of 7.0 to 9.5.[757]

In intensive culture systems, alkalinity should be maintained above 150 mg/ℓ (as $CaCO_3$) in order to maintain the buffering capacity of the water.[672] Crushed oyster shell is often used as a means of providing carbonate ions, but other sources, such as limestone or dolomite can be used.[847] Development of dense plankton blooms in ponds with low alkalinity can cause drastic diel shifts because CO_2 is removed during the day to support photosynthesis. As a result, the pH may rise to 10 or 11. At night, if planktonic respiration produces CO_2 in excess of the buffering capacity of the pond, the pH can drop to 6 or less. Addition of lime ($CaCO_3$) can help increase pond buffering capacity and reduce such drastic shifts.[841]

Hard water appears to benefit striped bass. Water with hardness of 25 to 30 mg/ℓ can lead to significant mortality unless salt is added.[846] Total hardness for striped bass culture should exceed 150 mg/ℓ, and 200 to 250 mg/ℓ is considered to be desirable.[757]

Responses of striped bass fry to both temperature and pH can be altered by increasing the level of dissolved solids.[848] Optimum conditions, calculated from a regression equation, appear to be 17.6°C, a pH of 7.5, and total dissolved solids at 185.7 mg/ℓ (as NaCl).

E. Nitrogen Compounds

High levels of ammonia, nitrite, and nitrate are more common problems in intensive culture systems than in ponds. Ammonia has been considered second only to low DO as a major cause of mortality in intensively cultured fish.[672] Bioassay studies have revealed that 96-hr LC_{50} values for NH_4OH in striped bass range from 1.5 to 2.8 mg/ℓ. Reduced feeding and growth, gill damage, and decreased disease resistance are among the sublethal signs of ammonia toxicity.[846] Ammonia concentration should not exceed 0.6 mg/ℓ in intensive culture systems.[757]

Nitrite and nitrate are not normally problems in pond culture unless intensive stocking or heavy fertilization is practiced. However, diel shifts in pH can alter the relative toxicity of nitrogen compounds and may sometimes lead to problems.[841] Little is known about the toxicity of nitrite and nitrate, though the latter is not generally considered to be toxic in the ranges which commonly occur even in intensive culture systems. Striped bass reportedly tolerate nitrate levels as high as 800 mg/ℓ, though larvae may be stressed when nitrate reaches 200 mg/ℓ.[757] Fingerlings can survive short-term exposure to nitrite concentrations of up to 1.4 mg/ℓ as NO_2-N (4.6 mg/ℓ NO_2^-).[672]

F. Turbidity and Light

Most streams where striped bass spawn can be characterized as turbid, but there is no evidence that turbidity is a requirement for successful spawning.[748] Turbidity levels greater than 1000 and 500 mg/ℓ have been reported as lethal to striped bass eggs and larvae, respectively,[849] but other studies indicated that egg hatch was not affected by levels as high as 5250 mg/ℓ, though larval development, was slowed, at turbidities over 1500 mg/ℓ.[850,851]

Direct sunlight adversely affects survival of 6.5-day-old SB X WB larvae.[852] The effects of increasing sunlight on reduced survival appear to be linear. Increased survival may be obtained by promoting turbidity or by shading ponds from above.[852]

VII. OTHER FACTORS AND PROBLEMS

A. Procurement of Brood Stock and Sperm Storage

Adult striped bass brood stock are normally collected on or near their spawning grounds. However, hatcheries sometimes have difficulty obtaining adequate supplies of ripe males.[757,772,797] In some locations, males are abundant early in the season but scarce later; in others, males are scarce throughout the spawning season. Techniques of cryopreservation of striped bass sperm have been developed on an experimental basis.[853] While fertilization rates are not normally as high as those obtained with fresh sperm, up to 87% fertilization has been obtained in smale-scale experiments and 60% on a production basis from cryopreserved sperm. More than 6 million larvae have been produced at the Moncks Corner, S.C. striped bass hatchery with cryopreserved sperm, and pond experiments have demonstrated that survival and growth of the fry compare favorably with those produced using fresh sperm. The techniques must be refined before they can be used with confidence for production.

B. Egg Quality

Poor egg quality occurs in various hatcheries. Egg quality is often lower early and late in the spawning season than during the peak, as evidenced by lower mean fertilization rates. Failure of eggs to mature within the females after hormone injection seems to be more frequent during the early and late portions of the spawning period as well, even though females appear to be eligible for induced ovulation. Although the time of ovulation is occasionally miscalculated, resulting in overripe eggs, the expertise at established hatcheries is such that this seldom occurs. With present technology, it is virtually impossible to predict egg quality until cell division begins and fertilization percentages are determined. As a result, semen that may be available in only limited quantities is frequently used on low quality eggs.

Various aberrations may also occur. Eggs are sometimes observed that, although they may have a high fertilization rate, have fragile, brittle chorions that burst at the slightest pressure. Prolarvae that hatch from those eggs appear normal but long-term survival is questionable.

Other aberrations, such as detachment of oil globules and developing cells from the yolk are sometimes seen. Because all of these phenomena are associated with eggs from specific females, they do not result from conditions in the hatchery. Whether such conditions are inherited or are the result of disease, pollutants, or other factors is not known; further research is clearly needed.

C. Stress, Handling, and Transport

Striped bass are more sensitive to handling stress than many warmwater species and considerable care is required to successfully hold and transport them.[672,757,765,776,782,784,854] Also, some life stages are more sensitive to stress than others.

Eggs and larvae are transported in sealed plastic bags with water and pure oxygen. The bags are placed in styrofoam containers to prevent rapid temperature changes. Eggs can be transported more successfully when they are incubated for at least 6 hr before being packed, and 12 to 16 hr of incubation provide the best results.[763] At the 12- to 16-hr stage, unfertilized eggs are opaque and more buoyant than live eggs and can be siphoned away. One recommendation calls for holding eggs for at least 24 hr prior to shipment with not more than 5000 eggs/ℓ of water.[772] Hatching during transport should be avoided because water fouling can result from the disintegration of chorions. Recommended temperatures during transport are 15.5 to 18.3°C.[772]

Larvae 1 to 2 days of age may survive transport better than older fish,[757,772] and one recommendation calls for shipping them at either 1 day or 5 days of age.[672] Larvae should not be exposed to bright light because of possible mortality.[757,782,855] For that reason, many agencies stock larval rearing ponds at night. In any case, larvae should be tempered with respect to temperature and other aspects of water quality before being introduced into culture systems.[672,757]

An apparently critical stage in development is that during transformation from postlarva to juvenile. Large numbers have reportedly died during that period.[672,794,831] The cause may be inadequate nutrition and failure of some postlarvae to convert to prepared feed.[672]

Harvesting and handling of fingerlings are critical to survival. Striped bass are more excitable than other commonly cultured species and fatal shock may occur, especially if there are additional stresses such as diseases, poor water quality, or sudden exposure to bright light or noise.[776,782,784,794]

Some stress of harvesting is apparently reduced if the operation occurs when pond temperatures are lowest and bright sunlight is not a factor. As a result many agencies draw down their ponds during the night and harvest at first light in the morning. Transfer of fish to cooler, more highly oxygenated water than that from which they were captured also seems helpful.

Handling stress also results in reduced resistance to disease.[757,794,854] Treatment of ponds with potassium permanganate has been recommended prior to harvest to reduce parasitic infestations.[782] Use of furacin and salt as prophylactic treatments following harvest and during hauling have also been recommended.[672,757,780] Anesthetics, such as quinaldine and MS-222, have been used with apparent success.[552,672,757,776] A combination of MS-222 and salt may reduce mortality associated with osmoregulatory dysfunction and resultant hypochloremia following handling and hauling stress in WB X SB hybrids.[552]

D. Swim Bladder Inflation

Failure of the gas bladder to inflate has been recognized as a potential source of striped bass mortality in recent years,[390,672,797,830,856] but information is lacking on the etiology and severity of the problem. The swim bladder normally inflates between the 5th and 7th day after hatching, simultaneous with initiation of feeding. Larvae appear to gulp atmospheric gas for initial inflation, and the percentage of larvae with inflated swim bladders is higher when fish are reared in containers with well-aerated, turbulent water.[390] However, histological work suggests that the mode of gas bladder inflation may be secretory in nature, but the mechanism is still not well understood.[856] Fish lacking swim bladders in culture ponds (termed "stargazers") have been observed swimming at or near the surface with their heads near the water surface.[858]

E. Disease

Most diseases that commonly afflict warmwater fish are also found in association with striped bass. Unnatural and stressful conditions that occur during culture encourage infections, and only hypoxia causes more mortality than disease in cultured fish.[757] Bacterial infections and protozoan parasites cause the most problems, though metazoan parasites such as trematodes (both monogenetic and digenetic), copepods, and acanthocephalans can also be harmful. Fungal infections are common following handling. Columnaris, caused by *Flexibacter columnaris*, is perhaps the most serious bacterial disease encountered by striped bass culturists. It occurs in ponds and intensive culture systems and is difficult to control.[757,776,777,796,805,857,858,859] Vibriosis may be a serious problem in marine systems, but can be prevented by vaccination.[822] Excellent descriptions of the common disease of striped bass, along with treatment recommendations are available.[757,850,859,860]

Results of bioassays on a number of therapeutic chemicals to assess their toxicity have been published,[757,782,861,862] as has the federal registration status of various treatment chemicals.[863] The effects of several treatment chemicals on biofilter efficiency have also been studied.[672]

VIII. THE FUTURE OF STRIPED BASS CULTURE

Striped bass and hybrid striped bass have excellent potential for continued and expanded recreational fisheries and for commercial culture. There is currently a market for fingerlings from both private and governmental sources.[821,864] At least seven commercial producers are now in operation and interest has been expressed by several other groups in the establishment of commercial ventures.

Much remains to be learned about these fishes, however. For example, a great deal more must be determined about the genetics of various striped bass stocks so that a desirable strain of domestic brood stock can be developed. Desirable characteristics would include rapid growth, good survival, disease resistance, and tolerance to a wide range of environmental conditions. Hybrid vigor has been demonstrated and should be researched in detail with respect to genetic manipulation of the parental species.

Increased production per unit area should be addressed. If fingerling production is to be increased in pond culture, an adequate supply of food must be provided and high water quality maintained. Several researchers are currently attempting to develop methods to promote production of desired zooplankton as well as increasing their production in ponds as a means of addressing the food availability problem. The Striped Bass Committee of the Southern Division, American Fisheries Society, considers the issue to be of high priority. Some progress has been made. For example, it has been reported that continuous aeration, combined with heavy fertilization and increased stocking densities, can lead to a 2.4-fold increase in fingerlings over that commonly obtained from conventional culture techniques.[792] Mechanical aeration during critical periods can provide sufficient oxygenation to support standing crops of at least 5500 kg/ha of SB X WB hybrids.

In view of recent advances in intensive culture, it would be advantageous to explore methods by which the date of larval production could be controlled. Environmental and physiological modifications may allow producers to adjust the timing of spawning to suit their needs. This could result in timing production with market demand. Progress has recently been achieved in advanced spawning of SB X WB hybrids by workers in South Carolina.[865]

ACKNOWLEDGMENTS

I thank Katrina Gray and Jeff M. Phillips for assistance in obtaining references and compiling data. J. M. Hinshaw and Deborah A. Zeleznick proofed the manuscript. The assistance of my co-authors, L. C. Woods III and M. T. Huish on a previous paper[825] is also gratefully acknowledged. M. T. Huish also reviewed the original manuscript and D. Wright typed it.

Financial support was provided in part by the Office of Sea Grant, NOAA, U.S. Department of Commerce under Grant Numbers NA79-AA-D-00048 and NA81-AA-D-00026, in part by the North Carolina Department of Administration, and in part by the U.S. Fish and Wildlife Service. The U.S. Government is authorized to produce and distribute reprints of this chapter for governmental purposes notwithstanding any copyright that may appear hereon.

The North Carolina Cooperative Fishery Research Unit is jointly sponsored by North Carolina State University, the North Carolina Wildlife Research Commission, and the U.S. Fish and Wildlife Service, Department of the Interior.

Chapter 10

BAITFISH

James T. Davis

TABLE OF CONTENTS

I. INTRODUCTION

At least 20 species of fishes have been raised on a commercial basis as bait.[869] In the U.S., only three warmwater species are of commercial importance. They are the golden shiner (*Notemigonus crysoleucas*), fathead minnow (*Pimephales promelas*), and goldfish (*Carassius auratus*). The latter species, in addition to being cultured for bait, is reared for the ornamental fish trade. Other species which are of local importance as bait include killifish (*Fundulus* spp.), chub suckers (*Erimyzon* spp.), stone rollers (*Campostoma* spp.), tilapia (*Tilapia* spp.), top minnows (*Poecilia* spp.), and shiners (*Notropis* spp.). The information presented here is confined to the first three species mentioned since they are currently the commercially important ones.

Baitfish production in the U.S. has increased from approximately 10,000 ha reported by Martin[869] in 1968 to over 22,500 ha reported by a Soil Conservation Service survey conducted in 1978. The golden shiner is captured in the wild but is primarily produced on fish farms in the midwest and south. The fathead minnow is reared in the midwestern U.S. in shallow lakes, but because of increased demand, the species is being intensively cultured in Arkansas, Louisiana, and Texas.

Goldfish, particularly the common variety, have been employed as bait for many years, and the production of the fancy varieties has escalated as demand in the aquarium trade has increased. There has also been an increased demand for goldfish as food items for predatory fish, particularly in the aquarium trade.

Certain aspects of the baitfish industry make it a highly competitive business. Bait is a perishable commodity which does not lend itself to stockpiling or the maintenance of large inventories. For that reason, competition is extremely keen in the marketing of bait to urban areas. Profits to bait producers can be quite high, even in the face of heavy competition.

II. GENERAL CULTURE CONSIDERATIONS

The life histories of the golden shiner and fathead minnows have been well documented,[870-872] as has that of the goldfish, both in the U.S. and Japan.[873,874] Culture ponds for baitfish are of the type generally utilized for the culture of other fishes[871,875] (refer also to Chapter 1). Special considerations in pond management which differ among the three species of interest are discussed below.

Water supply for a baitfish culture operation is even more critical than for other warmwater species. Under most conditions site selection should first take into account the water supply. Most operators prefer a large spring with a waterflow in excess of approximately 185 ℓ/min for each hectare of pond. Since there are few areas where spring water is available in the volumes necessary to support a large minnow farm, most bait producers construct their facilities in areas with plentiful supplies of cool well water. Artesian wells are preferred because of the reduced costs involved in pumping, but as is the case with flowing springs, sites with artesian water supplies are limited.

All of the species under discussion reach a peak in spawning intensity when the water temperature is above 20°C, so water at or above that temperature is most commonly sought by minnow and goldfish culturists. In addition, water of about 20°C is preferred for use in holding and hauling tanks. Therefore, one water temperature is adequate for all uses in most areas.

A problem with the use of well water involves the occasional presence of contaminants. Carbon dioxide can usually be removed by aeration prior to introduction of the water into culture chambers. Iron can be flocculated in the same manner. Iron-rich water introduced directly into holding tanks after areration should be sand filtered to remove the iron floc. This will reduce the chance of gill clogging in fish exposed to the suspended iron. If the water is introduced directly into ponds, sand filtration is often by-passed and the iron is allowed to settle *in situ.*

Surface waters such as streams or reservoirs are also used for baitfish culture under some circumstances. Water supplies of those types are generally considered to be much less satisfactory than springs or wells. Contamination from pesticides may occur, along with that present in runoff from feedlots, stables, municipalities, and domestic sewage effluents. In addition, most surface water supplies contain fish and other biological contaminants.

If surface water is to be used, it should be filtered through a sand filter or a screen material such as saran screen. The latter is relatively durable and permits more water flow than sand filters but it will allow the passage of many parasites. Also, close monitoring of screens is required so ruptures can be quickly discovered and screens replaced.

If undesirable fish are inadvertently admitted to culture ponds, they should be removed by draining the pond or by poisoning with a 5% rotenone solution applied at the rate of 2 mg/ℓ. Following rotenone treatment, a minimum of 2 weeks should be allowed for inactivation of the chemical prior to baitfish stocking.

Surface water is also often unsatisfactory because of high silt loads which are difficult and expensive to remove. Water from reservoirs can be a problem because of temperature. During the summer, reservoir surface water may be too warm, while deep water may be too cool and may be depleted in oxygen. Finally, surface waters are usually too cold during the winter to prevent ice formation on culture ponds, particularly in northern parts of the baitfish production region. Seining cannot be conducted when ponds are ice-covered.

Water from municipalities has been used in some instances but has not usually proved feasible for long periods because of the expense involved in chlorine removal. Chlorine can be effectively removed by charcoal filtration and the addition of sodium thiosulfate, but careful monitoring for residual chlorine is required when either method is employed.

Water pH has a decided effect on the efficacy of baitfish production as has been discussed by several authors.[870,872-874] Generally, baitfish can tolerate pH in the range of 6.0 to 9.0, but improved production has been obtained when pH is between 6.5 and 8.0. Control of pH is possible through the use of acid or alkaline-based fertilizers, but considerable expense is involved in their use.

In certain areas, water of low pH and concomitantly low alkalinity occur. The pH and alkalinity of such waters can be increased through the application of agricultural limestone, with the amounts used dependent upon soil and water analyses. For best production, both alkalinity and hardness should be between 50 and 300 mg/ℓ.

Dissolved oxygen (DO) requirements for baitfish are the same as for other species as discussed in Chapter 1; that is, the DO level should be maintained in excess of 3.0 mg/ℓ at all times. While all three species under discussion can survive lower levels, stress under reduced oxygen concentrations will result in reduced fish growth.

Salinity is of great importance in the culture of golden shiners. In waters of only 2 ppt salinity, production is severely curtailed. While goldfish tolerate 2 ppt, total production will be reduced.[876] Few data have been collected on the effect of salinity on fathead minnow production, but fish culturists report reduced production in water having salinities over 1.5 ppt.

III. REPRODUCTION AND GENETICS

The culture method that is to be used will determine much of the technology which can be applied to breeding. Golden shiners, fathead minnows, and goldfish can all be spawned in both extensive and intensive culture systems. Generally, when extensive systems are employed, the young fish are allowed to remain in the ponds with the adults. Harvest is usually by lift nets or traps, and those fish which are too small for sale are returned to allow them further growth. Pond sizes for extensive spawning and rearing range, in general, from 2 to 20 ha.

In intensive rearing systems, the eggs or fry are removed from spawning ponds and transferred to growout ponds. When the intensive system is utilized, spawning ponds of small size are employed. They are heavily stocked with broodfish which have been selected for maximum production.

Brood stock selection for fathead minnow production is usually done with a mechanical grader. Because the males are normally larger than the females, care should be taken to avoid selection of only the largest individuals for use as brood stock. In extensive culture, stocking rates of 1250 to 5000 adults per hectare are normally used. Higher stocking rates will usually result in the occurrence of stunted populations due to overproduction. In intensive culture, broodfish stocking densities from 35,000 to 60,000 are commonly used, though research has demonstrated that 50,000 adults at the ratio of 5 females:1 male results in best production.[872]

Little organized effort has been devoted to selection of improved strains of fathead minnows. In part this is due to the short lifespan of the species, but it is also an indication that most culturists are satisfied with the results being obtained.

Golden shiners are also grown in both extensive and intensive systems. Stocking rates for ponds used in extensive culture are generally about 5000 fish/ha and the fish selected are usually of mixed sizes (designated "pond run") to avoid the possibility of skewed sex ratios which could occur with mechanical grading since golden shiner females grow more rapidly than the males. When intensive culture is employed with egg transfer, broodfish stocking densities of 400 to 500 kg/ha are common. The sex ratio for those broodfish is also "pond run".

Methods employed for spawning goldfish are much the same as those for golden shiners. The number of broodfish stocked depends on their size, which directly affects the number of eggs produced per female. For extensive systems, from 250 to 750 broodfish are stocked per hectare, while 800 to 1000 kg/ha are stocked in intensive culture.

The genetics of goldfish have been studied in detail and there have been many varieties developed, primarily for the aquarium trade. Variety development has been the result of selective breeding and some cross-breeding. Some of the more common goldfish breeds are the lionhead, comet, shubunkin, calico, and common. Procedures for breeding the varieties are usually passed from culturist to culturist, though some of the methodology has been published.[874]

IV. POND MANAGEMENT AND SPAWNING PROCEDURES

A. Fathead Minnows

Fathead minnows have the unique habit of spawning on the underside of vegetation or other objects found in the water. Therefore, aquatic plants are commonly allowed to remain in culture ponds when the extensive technique is utilized. There is a danger in leaving too many plants, however, since oxygen depletions may occur from decaying plant material. Utilization of plants as natural spawning substrates is widely employed, but has the disadvantage that overpopulation and stunting sometimes occur. Because most of the broodfish die after spawning, there is little interference between adults and their offspring. In at least some culture installations, it is believed that the older fingerlings compete with fry for food. That situation may explain some of the variable production levels which have been reported.

In intensive culture, the fry transfer method is commonly employed. To increase spawning area, rock, pieces of bricks, and boards are used to supplement existing spawning sites. Old lumber is often stapled to wires stretched parallel across the pond. Boards should be provided at the rate of 41 cm of 2.5 × 10 cm lumber per 100 male broodfish. Some culturists furnish cubicles for the males to use for territories during nest protection, but that technique has not led to increased fry production.[872]

Egg incubation requires 5 to 7 days depending on water temperature. After absorption of their yolk sacs, the fry feed on phytoplankton for about 2 weeks after which they will readily consume prepared diets. During the first 2-week period of feeding, the fry are usually captured in lift nets or seines and transferred to rearing ponds. Some data are available which indicate that more fry can be produced from the same area of spawning ponds if the ponds are drained and all fry removed at least twice during each growing season.[877] Other methods are utilized in the production of fathead minnows for forage in hatchery situations.[878-880]

When rearing ponds are employed, the fry are directly stocked from spawning ponds. The rearing ponds are fertilized in advance of stocking and receive fry at the rate of 125,000 to 750,000/ha. The accepted practice is to count the number of fish in a kilogram (determined by weight added to a previously tared container) and then calculate the total weight required to reach the desired stocking density for each pond. The number of fry in a kilogram may be rather high, so 0.1-kg sample counts are sometimes employed.

Fish should be weighed in a small amount of water and handled carefully to avoid stress. As a general rule, stocking numbers make allowance for 25 to 50% mortality during the growing season. To reach salable size within a single growing season, about 250,000 fish/ha is the maximum stocking rate. If all or a portion of the second growing season is to be used to produce marketable fish, two or three times as many minnows can be reared within the growout ponds.[872] Overwinter survival of fish of a size equivalent to over 11,000 fish/kg is usually quite high, but the postspawning mortality of broodfish often exceeds 80%.

B. Golden Shiners and Goldfish

Since the methods used for production of golden shiners and goldfish are virtually the same, the two species will be discussed together. In the extensive culture or free spawning method, pond water level is lowered during the late winter or early spring to stimulate the growth of grass along the shoreline. In many instances rye grass is planted in that area.

When the ponds are refilled, the plants become spawning sites. Aquatic plants will also serve that purpose, as will rice. Care must be exercised when using the latter, however, as it may become so dense so to preclude effective management of the pond. If plant growth is inadequate, on the other hand, hay or straw may be placed near the shoreline as spawning substrate. When that method is used, the material should be staked down to prevent it from piling up at one end of the pond in response to wind generated movement.

Spawning activity may decline for a variety of reasons, but is usually dependent upon water quality conditions. The usual method of promoting active spawning is to add cool water rapidly and raise the water level in the pond. The technique has also been employed to extend the spawning season in small ponds. Other methods of inducing spawning include the use of chemicals to "shock" the fish. While such methods are commonly employed, there is little literature on the subject.[870] If possible, the broodfish should be removed from the pond as they begin to compete with young-of-the-year fish. Many culturists either retain them for reuse as broodfish or sell them as bait.

Intensive culture of golden shiners and goldfish is usually practiced using the egg transfer method. When that technique is employed, it is essential to remove all natural vegetation from spawning ponds so that egg deposition can be controlled. Research indicates that the presence of aquatic vegetation as well as certain temperatures and photoperiods can cause spontaneous ovulation.[881] Marginal aquatic plants must not be allowed in the pond. Soil sterilants are often applied around pond margins.

When the fish are ready to spawn, mats are placed in the pond. Their location is often prepared in advance using gravel to level the area where the mats are to be placed. Spawning mats are approximately 30 cm × 60 cm and are constructed by sandwiching Spanish moss between 10 cm × 10 cm mesh pieces of welded wire. The tops and bottoms of each frame are wired or hog-ringed together.

Sometimes it is necessary to stimulate spawning by rapidly raising the water level, adding fresh, cool water or by "chemical shock". It is important to place spawning mats 2.5 cm below the water surface and to put them on level substrates. In most cases several mats are placed end-to-end parallel to the shoreline. The mats are ready for transfer when they are uniformly covered with eggs. Allowing too many eggs on a given mat encourages fungus growth which will reduce the hatching rate.

The number of mats which are placed in a pond and length of time between placement and recovery will depend on the spawning activity of the broodfish. Goldfish usually spawn just after dawn with spawning continuing until the sun strikes the pond, while golden shiners may spawn throughout the day.

Mat densities in spawning ponds should not exceed the number which can be covered with eggs within a given spawning period. If the fish spawn more rapidly than expected, filled mats can be removed and replaced with fresh ones or new mats can be stacked on top of filled ones to prevent egg predation. Such predation is a constant problem in goldfish production ponds. Too many mats in the pond will result in underutilization.[870] At the same time, too few mats may cause the fish to use alternative sites for spawning.

When the mats are filled, they are moved to rearing ponds. Estimation of egg numbers on mats is usually a function of experience. If the mats are well filled, from 125 to 200 mats are placed in each hectare of rearing pond for golden shiners, though as many as 400 mats are often used per hectare in goldfish ponds. It is common practice to leave the mats in the rearing ponds for 10 days, after which they are removed and thoroughly washed. They can then be reintroduced into spawning ponds as needed.

Fry transfer is also practiced with golden shiners and goldfish because fry numbers can readily be determined and rigid control over stocking rates is possible. Determination of fry numbers is accomplished utilizing the weight-count method outlined above or by counting the number of fry in a 30-mℓ sample and calculating the volume displacement of fish required to stock a given pond.

The number of fish stocked per hectare depends upon the time available between stocking and marketing, size of fish desired, level of management employed, and length of the growing season. For golden shiners, stocking rates of 125,000 to 500,000/ha may be used, while goldfish are usually stocked at from 50,000 to 2,500,000/ha. The high stocking rates are used in instances where fish are stockpiled in ponds overwinter for growout the following season. The fish are restocked at reduced densities in the spring and can be expected to grow rapidly to market size.

V. FEEDING AND NUTRITION

The food available for baitfish fry depends to a large extent on the abundance of plankton available in spawning and rearing ponds. Preparation for fish usually involves allowing ponds to dry for several days after which they are rapidly filled with water. Liquid fertilizer is added as the water enters the ponds. In some instances, plankton from an adjacent pond is seeded into newly filled ponds. Within 4 or 5 days, the spawning mats are placed into newly filled spawning ponds. When the eggs hatch about 5 days later, there is a ready supply of food available to the fry. For fathead minnows, where spawning mats are not employed, pond preparation begins about 10 days prior to stocking.

In most ponds, the water is fertilized at a rate sufficient to produce a Secchi disc reading of about 20 cm. That level of fertility is maintained by intermittent addition of fertilizer until the fish reach about 2.5 cm, after which the bloom is allowed to decline.

As soon as the fry begin to come to the water surface they should be offered prepared feed. Feed should be made available a minimum of twice daily, and many growers feed four to eight times daily. The normal feeding method is to provide finely ground feed that will float. The feed is distributed along the upwind side of the pond, from which it will spread over the pond surface. Sufficient feed should be applied to ensure that it is available to all parts of the pond.

During the initial stages of feeding, total consumption may not occur. Nonetheless, feeding should be continued, even if the fish are being overfed, since it is important that the fish receive ample feed whenever they are willing to eat. After the fish are eating on a regular basis, twice daily feeding is sufficient.

As the fish increase in size, the feed should be gradually changed from a finely ground product to one which is more coarsely ground. Subsequently, crumble-sized particles may be fed, and finally, the fish can be offered pelleted feed. When a change from one particle size to another is made, that change should be gradual. One part of the new particle size should be mixed with four parts of the previous feed size until the fish are observed consuming the new particle size. Thereafter, the ratio of new particle size to old can be gradually increased daily until only the new size is fed. Most fish appear able to detect changes in feed ingredients and may ignore new formulations for a period of time.

The amount of feed to be offered baitfish in ponds depends to a large extent on what the fish will readily consume and the time available between onset of feeding and marketing. For most rapid growth, the fish should be fed all they will consume in 2 hr twice daily. The amount can be reduced if less than maximum growth is desired.

The nutritional requirements of baitfish have not been extensively studied.[870] It is generally assumed that it will be necessary to balance any prepared feed with natural food. Therefore, it is essential to encourage plankton blooms in production ponds. For most growers, this is best accomplished with a combination of organic and inorganic fertilizers. Depending upon pond condition, from 1000 to 2500 kg/ha of manure and 30 ℓ of inorganic liquid fertilizer (11-37-0) are applied with the incoming water. Additional liquid fertilizer is added as needed.

It is generally accepted that the production of fish can be doubled through the addition of prepared feeds, so such feeds are widely employed by baitfish producers. Fry feeds are generally higher in protein (38%) than grower feeds (30 to 35%). Formulas are available from several sources which comply with the protein and particle size standards which have been adopted by the industry.[870,871]

Because goldfish are usually more valuable than other baitfish species and it is essential to get them on prepared feeds quickly to promote rapid growth, many growers add hard-boiled egg yolks (filtered through cotton cloths) or commercial grade egg yolk. These may be substituted for other ingredients in starter rations or they may be fed alone for the first 2 to 3 days after the fry begin to feed.[871,874]

One of the main recommendations on feeding minnows and goldfish is that the fish, except as newly feeding fry, should not be overfed. After the first several days of feeding the rate should be reduced if the ration is not completely consumed. When the weather is very hot the fish should be fed during the coolest part of the day. If the air temperature is above 30°C or the sky is overcast, the amount of feed offered should be reduced. Most producers limit the amount of feed made available to the fish to 30 kg/ha/day during the summer.

VI. VEGETATION CONTROL

The best method of vegetation control is prevention. Most production ponds have few problems with aquatic vegetation and control is effected by closely mowing the grass near pond edges to discourage predators. If problems develop, various methods of vegetation control are available.[882]

For intensive culture, complete vegetation removal from spawning ponds is required. Chemical control is often supplemented by mechanical removal by hand.

VII. HARVESTING AND HANDLING

Harvesting of minnows, especially shiners, during the summer months requires a greal deal of patience and attention to detail. Any time the air temperature is above 25°C or the surface water temperature is above 22°C, the movement of fish should be confined to the coolest parts of the day. This usually means that if seining is to be conducted, all the work should be accomplished prior to 1000 hr. If the fish are being captured with lift nets, the work can continue until later in the day if the water in the hauling tank has been cooled to or below 18°C and is well oxygenated.

When seining, care should be taken to avoid capturing more fish than are required. Seining the entire pond to capture a small number of fish will often result in heavy stress and subsequent disease epizootics may develop. As the seine is landed, it is usually shortened until it can be bagged and then moved out from shore to deeper water where the fish will not be so crowded. Then the seine is staked out with steel rods.

During crowding, the fish undergo a period of excitement and severe stress. The fish tend to brush against one another and against the net, causing the loss of scales. Damage to the fish provides not only an avenue for entrance of disease organisms, but may also cause sufficient mutilation that the fish are no longer salable. Under certain circumstances of extreme crowding the fish may become so excited that suffocation results. It is generally recommended that no more than 250 kg of fish be seined at one time so overcrowding can be avoided.

Fish are removed from the seine with a long-handled dip net made of soft nylon with a mesh size of about 5 mm. Nets which will not hold more than about 1.5 kg are recommended. The fish are carefully placed in buckets partially filled with 18°C water from the hauling tank. The buckets are moved directly to the hauling tank and the fish gently introduced into the tank.

Seines for use in harvesting minnows and goldfish have been discussed by other authors and the technology is well documented.[869-872] Perhaps the most important element is that the seine should be of soft material so damage to the fish can be reduced. The same condition holds for other nets utilized in the capture of baitfish.

Lift nets are often used to harvest minnows and goldfish. These nets are usually fastened by lines to a long pole from which the nets are lowered into the water. Feed is used to attract the fish over the net. When the operator perceives that sufficient numbers of fish are present, the net is rapidly lifted and the fish are captured. The procedure of removing the fish from the bag of the lift net is the same as that described for seines.

There are many designs for hauling tanks utilized to carry fish from ponds to holding facilities.[869-872] Actual design is dependent on the quantity of fish to be hauled and the distance of the trip. Holding tank water temperature should be held below 22°C and the tanks should not be overloaded. The major criterion for short hauls is that no more than 1 kg of fish should be placed in each 10 ℓ of water during hot weather. If the trip from the pond bank to the holding facility is of more than 20-min duration, the weight of fish in the hauling tank should be reduced by 10% for each additional 20 min.

In general, 1 kg of fish per 10 ℓ of water will hold minnows for up to a week if the water is well aerated. Normally, one agitator is recommended for each 50 kg of fish in a holding tank. The agitator should be of at least 1/25 hp and should operate continuously. In addition, fresh water of the same temperature as that in the holding tank, or less (down to 18°C) should be added to the tank at least once daily. Most culturists flush their tanks with a complete change of water after the first 6 hr and at least daily thereafter.

Most minnows and feeder goldfish are graded prior to sale. Mechanical graders are used for this procedure and are available in a variety of shapes and sizes. Generally, grader bars are made with a desired spacing between them. Fish below a given size pass through each grader, while larger fish are retained. Retention is affected by the plumpness of the fish, the species, and other factors. After the fish are graded, they are counted to determine the number per thousand as that is the measure under which the fish are marketed. Details of the grading procedure have been documented in the literature.[869,870]

Following harvest, most fish are held without handling for at least 24 hr. During that period the fish are kept in cool water and recover from the stress to which they were exposed during harvesting. The recovery period enables them to better withstand subsequent handling. An additional advantage of the delay before secondary handling is that the fish have an opportunity to clear their intestinal tracts of fecal material. This helps prevent water quality deterioration during the grading process.

Holding facilities are where most mortalities of minnows and goldfish are documented, therefore special care should be taken to minimize mortalities during holding. Details of recommended practices aimed at reducing holding mortalities have been discussed by Martin[869] and Johnson.[883]

Movement from holding tanks to retail establishments requires the same type of hauling equipment utilized to transport the fish from ponds to holding facilities. For long distance haulding, larger trucks are often utilized since they can carry more fish than small trucks. Alternatively, fish can be shipped in plastic bags charged with oxygen as discussed by Harry.[884] One of the primary considerations is that the fish should not be exposed to rapid changes in temperature during handling and hauling.

VIII. DISEASES AND PREDATORS

Bacterial and fungus infections may kill large numbers of baitfish but are usually associated with stress and injury which occur during handling. Those problems may also occur following environmental degradation.

Most of the losses in baitfish ponds are attributable to parasites, the major ones of which are protozoans. Many parasites are common to both minnows and goldfish and can cause extensive losses. The major external parasites are *Trichodina, Ichthyobodo, Chilodonella, Cryptobia*, and *Ichthyophthirius*. Most are relatively easy to recognize with the aid of a microscope. Control and management recommendations have been outlined by Giudice et al.[870] and Bishop.[885]

Sporozoan parasites are among the most difficult to control. The major ones are *Mitraspora cyprini* in goldfish, *Myxobolus notemigoni* (milk scale disease) in golden shiners, and *M. argenteus* in fathead minnows.[870,885] *Pleistophora ovariae* affects the ovaries of golden shiners and may adversely affect production in that species. Control is best achieved by utilizing 1-year-old brookfish.

Of the internal parasites, the flukes are the most damaging to fish populations. Control is dependent to a large extent on good management practices as reviewed by Giudice.[871] Tapeworms and round worms are a problem in certain areas but are not generally found in sufficient numbers to cause high mortalities among baitfish. Crustacean parasites are a recurring problem, however. In the goldfish industry, the fish louse *Argulus* is a major cause of losses in some hatcheries. The anchor parasite is a major source of losses in the other two bait species.

Predators are a major problem for all baitfish producers. Predation occurs at all life stages of the fish. Fish, birds, frogs, snakes, turtles, insects, and alligators all impact upon baitfish populations. Control of each of these requires careful management, and there is a further danger with some predators because they may be vectors of diseases. Control is usually achieved by physical removal of the offending species. This can be costly, but in many instances can make the difference between profit and loss. More information is available in publications from various state fish and game agencies.[871]

IX. MARKETING

Marketing represents a discipline in which little information is available. It is an acknowledged fact that there are many outlets for baitfish, but knowledge on how to sell to wholesalers and retailers is usually discovered by the individual grower working alone. Pricing is usually dependent upon demand and the extent to which competition exists within a particular area. In regions where there are numerous suppliers competing for the same market, heavy reliance is placed on the services supplied to the buyer. For example, a minnow supplier may be required to visit a large retailer on a tightly fixed schedule and may also be expected to deliver other types of bait (e.g., worms and crickets) as well as fishing tackle to the retailer.[886] In other areas, service is secondary to price and/or supply. Most wholesaling is done from trucks while retailing is often done using plastic bags or by placing the baitfish directly into minnow buckets furnished to the buyer.

The future of the baitfish industry seems bright in that each year there is an increase in the number of fishing licenses sold and many fishermen are convinced that the best way to assure a good catch is by using live bait. There seems to be ample marketing opportunity for the large operator who can haul fish for long distances and dependably supply holding facilities. At the same time, there is room in the industry for small producers who develop local markets and furnish high quality baitfish to the public.

REFERENCES

1. Stickney, R. R., *Principles of Warmwater Aquaculture,* Interscience, New York, 1979, 1.
2. Robins, C. R., Bailey, R. M., Bond, C. E., Brooker, J. R., Lachner, E. A., Lea, R. N., and Scott, W. B., *A List of Common and Scientific Names of Fishes from the United States and Canada,* American Fisheries Society, Bethesda, Maryland, 1980, 1.
3. Trewavas, E., Tilapias: taxonomy and speciation, in *The Biology and Culture of Tilapia,* Pullin, R. S. V. and Lowe-McConnell, R. H., Eds., International Center for Living Aquatic Resources Management, Manila, Philippines, 1982, 3.
4. Wheaton, F. W., *Aquaculture Engineering,* Interscience, New York, 1977, 1.
5. Allen, L. J. and Kinney, E. C., Eds., *Proc. of the Bio-Engineering Symp. for Fish Culture,* Fish Culture Section, American Fisheries Society, Washington, D.C., 1981, 1.
6. Boyd, C. E., Water Quality in Warmwater Fish Ponds, Agricultural Experiment Station, Auburn, Ala., 1979, 1.
7. Arce, R. G. and Boyd, C. E., Effects of agricultural limestone on water chemistry, phytoplankton productivity, and fish production in soft water ponds, *Trans. Am. Fish. Soc.,* 104, 308, 1975.
8. Boyd, C. E. and Walley, W. W., Total Alkalinity and Hardness of Surface Waters in Albama and Mississippi, Agricultural Experiment Station, Auburn, Ala., 1975, 1.
9. Hochachaka, P. W., Intermediary metabolism in fishes, in *Fish Physiology,* Vol. I, Hoar, W. S. and Randall, D. J., Eds., Academic Press, New York, 1969, 351.
10. Chipman, W. A., Jr., The Role of pH in Determining the Toxicity of Ammonium Compounds, Ph.D. dissertation, University of Missouri, Columbia, 1934, 1.
11. Wuhrmann, K. and Woker, H., Experimentelle Untersuchungen uber die Ammoniak-und Blausaurevergiftung, *Schweiz. Z. Hydriol.,* 11, 210, 1948.
12. Wuhrmann, K., Zehender, F., and Woker, H., Uber die fischereibiologische Bedeutung des Ammonium- und Ammiakgehaltes fliessender Gewasser, *Vierteljahrsschr. Naturforsch. Ges. Zur.,* 92, 198, 1947.
13. Emerson, K., Russo, R. C., Lund, R. E., and Thurston, R. V., Aqueous ammonia equilibrium calculations: effect of pH and temperature, *J. Fish. Res. Bd. Can.,* 32, 2379, 1975.
14. Redner, B. D. and Stickney, R. R., Acclimation to ammonia by *Tilapia aurea, Trans. Am. Fish. Soc.,* 108, 383, 1979.
15. Konikoff, M., Toxicity of nitrite to channel catfish, *Prog. Fish-Cult.,* 37, 96, 1975.
16. Tomasso, J. R., Wright, M. I., Simco, B. A., and Davis, K. B., Inhibition of nitrite-induced toxicity in channel catfish by calcium chloride and sodium chloride, *Prog. Fish-Cult.,* 42, 144, 1980.
17. Boyd, C. E., Romaire, R. P., and Johnston, E., Predicting early morning dissolved oxygen concentrations in channel catfish ponds, *Trans. Am. Fish Soc.,* 107, 484, 1978.
18. Boyd, C. E. and Tucker, C. S., Emergency aeration of fish ponds, *Trans. Am. Fish. Soc.,* 108, 299, 1979.
19. Hollerman, W. D. and Boyd, C. E., Nightly aeration to increase production of channel catfish, *Trans. Am. Fish. Soc.,* 109, 446, 1980.
20. Tang, Y. A., Physical problems in fish farm construction, in *Advances in Aquaculture,* Pillay, T. V. R. and Dill, W. A., Eds., Fishing News Books Ltd., Surrey, England, 1979, 99.
21. National Research Council, Nutrient Requirements of Warmwater Fishes and Shellfishes, National Academy of Science Press, Washington, D.C., 1983, 1.
22. Fouche, L., Webb, P., Killcreas, W., Waldrop, J. E., Chin, K., Kennedy, N., and Paschal, R., A Records System for Catfish Production Management Decision-Making . . . for These Microcomputers: Radio Shack Model III Radio Shack Model II Radio Shack Model 16 IBM Personal Computer, Agricultural Economics Tech. Publ. No. 43, Mississippi Agricultural and Forestry Experiment Station, Mississippi State, 1983, 1.
23. Hickel, R., Killcreas, W. E., and Waldrop, J. E., A Catfish Growth Stimulation Model: for Use with TRS-80 Models II, III, 16, and IBM PC Microcomputers, Agricultural Economics Tech. Publ. No. 42, Mississippi Agricultural and Forestry Experiment Station, Mississippi State, 1983, 1.
24. Swingle, H. S., Preliminary results of the commercial production of channel catfish in ponds, *Proc. S. E. Assoc. Game Fish Comm.,* 10, 160, 1956.
25. Swingle, H. S., Experiments on growing fingerling channel catfish to marketable size in ponds, *Proc. S.E. Assoc. Game Fish Comm.,* 12, 63, 1958.
26. Brown, E. E., *World Fish Farming Cultivation and Economics,* AVI Publishing, Westport, Conn., 1977, 1.
27. Wellborn, T. L. and Schwedler, T. E., For Fish Farmers, Newsletter from the Cooperative Extension Service, Mississippi State University, January 15, 1982, 1.
28. Wellborn, T. L., personal communication.
29. Reagen, R. E., Jr., Survey of current practices and problems in the Mississippi catfish industry, *Proc. S.E. Assoc. Game Fish Comm.,* 34, 127, 1980.

30. Waldrop, J. E., personal communication.
31. Parker, N. C. and Simco, B. A., Evaluation of recirculating systems for the culture of channel catfish, *Proc. S.E. Assoc. Game Fish Comm.,* 27, 474, 1974.
32. Broussard, M. C., Jr. and Simco, B. A., High density culture of channel catfish in a recirculating system, *Prog. Fish-Cult.,* 38, 138, 1976.
33. Lewis, W. M. and Buynak, G. L., Evaluation of a revolving plate type biofilter for use in recirculated fish production and holding units, *Trans. Am. Fish. Soc.,* 105, 704, 1976.
34. Broussard, M. C., Jr., Parker, N. C., and Simco, B. A., Culture of channel catfish in a high flow recirculating system, *Proc. S.E. Assoc. Game Fish Comm.,* 27, 745, 1974.
35. Lewis, W. M., Yopp, J. H., Schramm, H. L., Jr., and Brandenburg, A. M., Use of hydroponics to maintain quality of recirculated water in a fish culture system, *Trans. Am. Fish. Soc.,* 107, 92, 1978.
36. Sutton, R. J. and Lewis, W. M., Further observations on a fish production system that incorporates hydroponically grown plants, *Prog. Fish-Cult.,* 44, 55, 1982.
37. Andrews, J. W., Knight, L. H., Page, J. W., Matsuda, Y., and Brown, E. E., Interactions of stocking density and water turnover on growth and food conversion of channel catfish, *Prog. Fish-Cult.,* 33, 197, 1971.
38. Tilton, J. E. and Kelley, J. E., Experimental cage culture of channel catfish in heated discharge water, *Proc. World Maricult. Soc.,* 1, 73, 1970.
39. Pennington, C. H., Cage Culture of Channel Catfish, *Ictalurus punctatus* (Rafinesque), in a Thermally Modified Texas Reservoir, Ph.D. dissertation, Texas A&M University, College Station, 1977, 1.
40. Schmittou, H. R., Developments in the culture of channel catfish, *Ictalurus punctatus* (Rafinesque), in cages suspended in ponds, *Proc. S.E. Assoc. Game Fish Comm.,* 23, 226, 1969.
41. Ray, L., Channel catfish production in geothermal water, in Proc. Bio-Eng. Symp. Fish Culture, Travers City, Mich., Fish Culture Section, American Fisheries Society, 1981, 192.
42. Hill, T. K., Chesness, J. L., and Brown, E. E., Growing channel catfish, *Ictalurus punctatus* (Rafinesque), in raceways, *Proc. S.E. Assoc. Game Fish Comm.,* 27, 488, 1974.
43. Tuten, J. S. and Avault, J. W., Jr., Growing red swamp crayfish (*Procambarus clarkii*) and several North American fish species together, *Prog. Fish-Cult.,* 43, 97, 1981.
44. Clady, M. D., Cool-weather growth of channel catfish held in pens alone and with other species, *Prog. Fish-Cult.,* 43, 92, 1981.
45. Wilson, J. L. and Hilton, L. R., Effects of tilapia densities on growth of channel catfish, *Prog. Fish-Cult.,* 44, 207, 1982.
46. Huner, J. V., Avault, J. W., Jr., and Bean, R. A., Interactions of freshwater prawns, channel catfish fingerlings, and crayfish in earthen ponds, *Prog. Fish-Cult.,* 45, 36, 1983.
47. Perry, W. G., Jr. and Avault, J. W., Jr., Comparisons of striped mullet and tilapia for added production in caged catfish studies, *Prog. Fish-Cult.,* 34, 229, 1972.
48. Perry, W. G., Jr. and Avault, J. W., Jr., Polyculture studies with channel catfish and buffalo, *Proc. S.E. Assoc. Game Fish Comm.,* 29, 91, 1975.
49. Merkowsky, A. and Avault, J. W., Jr., Polyculture of channel catfish and hybrid grass carp, *Prog. Fish-Cult.,* 38, 76, 1976.
50. Williamson, J. and Smitherman, R. O., Food habits of hybrid buffalofish, tilapia, Israeli carp and channel catfish in polyculture, *Proc. S.E. Assoc. Game Fish Comm.,* 29, 86, 1975.
51. Perry, W. G., Jr. and Avault, J. W., Jr., Polyculture studies with blue, white and channel catfish in brackish water ponds, *Proc. S.E. Assoc. Game Fish Comm.,* 25, 466, 1972.
52. Green, O. L., Comparison of production and survival of channel catfish stocked alone and in combination with blue and white catfish, *Prog. Fish-Cult.,* 35, 225, 1973.
53. Lewis, W. M., Observations on the grass carp in ponds containing fingerling channel catfish and hybrid sunfish, *Trans. Am. Fish. Soc.,* 107, 153, 1978.
54. Kilgen, R. H., Growth of channel catfish and striped bass in small ponds stocked with grass carp and water hyacinths, *Trans. Am. Fish. Soc.,* 107, 176, 1978.
55. Gabel, S. J., Avault, J. W., Jr., and Romaire, R. P., Polyculture of channel catfish (*Ictalurus punctatus*) with all-male tilapia hybrids (*Sarotherodon mossambica* male x *Sarotherodon hornorum* female), *J. World Maricult. Soc.,* 12, 153, 1981.
56. Stickney, R. R., Rearing channel catfish fingerlings under intensive culture conditions, *Prog. Fish-Cult.,* 34, 100, 1972.
57. Murai, T., High-density rearing of channel catfish fry in shallow troughs, *Proc. Fish-Cult.,* 41, 57, 1979.
58. Lewis, W. M., Cage culture of channel catfish, *Catfish Farmer,* 1(4), 5ff, 1969.
59. Kilambi, R. V., Adams, J. C., Brown, A. V., and Wickizer, W. A., Effects of stocking density and cage size on growth, feed conversion, and production of rainbow trout and channel catfish, *Prog. Fish-Cult.,* 39, 62, 1977.

60. Allen, K. O., Effects of stocking density and water exchange rate on growth and survival of channel catfish *Ictalurus punctatus* (Rafinesque) in circular tanks, *Aquaculture*, 4, 29, 1974.
61. Kilambi, R. W., Noble, J., and Hoffman, C. E., Influence of temperature and photoperiod on growth, food consumption and food conversion efficiency of channel catfish, *Proc. S.E. Assoc. Game Fish Comm.*, 24, 519, 1970.
62. Andrews, J. W. and Stickney, R. R., Interactions of feeding rates and environmental temperature on growth, food conversion, and body composition of channel catfish, *Trans. Am. Fish. Soc.*, 101, 94, 1972.
63. Andrews, J. W., Knight, L. H., and Murai, T., Temperature requirements for high density rearing of channel catfish from fingerling to market size, *Prog. Fish-Cult.*, 34, 240, 1972.
64. Carlson, A. R., Blocher, J., and Herman, L. J., Growth and survival of channel catfish and yellow perch exposed to lowered constant and diurnally fluctuating dissolved oxygen concentrations, *Prog. Fish-Cult.*, 42, 73, 1980.
65. Allen, K. O., Effects of flow rate and aeration on survival and growth of channel catfish in tanks, *Prog. Fish-Cult.*, 38, 204, 1976.
66. Andrews, J. W., Murai, T., and Gibbons, G., The influence of dissolved oxygen on the growth of channel catfish, *Trans. Am. Fish. Soc.*, 102, 835, 1973.
67. Dunham, R. A., Smitherman, R. O., and Webber, C., Relative tolerance of channel x blue hybrid and channel catfish to low oxygen concentrations, *Prog. Fish-Cult.*, 45, 55, 1983.
68. Boyd, C. E. and Tucker, C. S., Emergency aeration of fish ponds, *Trans. Am. Fish. Soc.*, 108, 299, 1979.
69. Loyacano, H. A., Effects of aeration in earthen ponds on water quality and production of white catfish, *Aquaculture*, 3, 261, 1974.
70. Hollerman, W. D. and Boyd, C. E., Nightly aeration to increase production of channel catfish, *Trans. Am. Fish. Soc.*, 109, 446, 1980.
71. Tucker, C. S. and Boyd, C. E., Effects of simazine treatment on channel catfish and bluegill production in ponds, *Aquaculture*, 15, 345, 1978.
72. Romaire, R. P. and Boyd, C. E., Effects of solar radiation on the dynamics of dissolved oxygen in channel catfish ponds, *Trans. Am. Fish. Soc.*, 108, 473, 1979.
73. Robinette, H. R., Effect of selected sublethal levels of ammonia on the growth of channel catfish, *Prog. Fish-Cult.*, 38, 26, 1976.
74. Colt, J. and Tchobanoglous, G., Evaluation of the short-term toxicity of nitrogenous compounds to channel catfish, *Ictalurus punctatus*, *Aquaculture*, 8, 209, 1976.
75. Knepp, G. L. and Arkin, G. F., Ammonia toxicity levels and nitrate tolerance of channel catfish, *Prog. Fish-Cult.*, 35, 221, 1973.
76. Tomasso, J. R., Goudie, C. A., Simco, B. A., and Davis, K. B., Effects of environmental pH and calcium on ammonia toxicity in channel catfish (*Ictalurus punctatus*), *Trans. Am. Fish. Soc.*, 109, 229, 1980.
77. Mitchell, S. J. and Cech, J. J., Is ammonia a direct causative agent for gill damage in channel catfish? Compounding effects of residual chlorine in the test water (Abstr.), in 113th Annu. Meet. Am. Fish. Soc., University of Wisconsin, Milwaukee, August 16 to 20, 1983.
78. Colt, J. and Tchobanoglous, G., Chronic exposure of channel catfish, *Ictalurus punctatus*, to ammonia: effects on growth and survival, *Aquaculture*, 15, 353, 1978.
79. Tomasso, J. R., Simco, B. A., and Davis, K. B., Inhibition of ammonia and nitrite toxicity to channel catfish, *Proc. S.E. Assoc. Fish. Wildl. Agen.*, 33, 600, 1979.
80. Worsham, R. L., Nitrogen and phosphorus levels in water associated with a channel catfish (*Ictalurus punctatus*) feeding operation, *Trans. Am. Fish. Soc.*, 104, 811, 1975.
81. Konikoff, M., Toxicity of nitrite to channel catfish, *Prog. Fish-Cult.*, 37, 96, 1975.
82. Tomasso, J. R., Simco, B. A., and Davis, K. B., Chloride inhibition of nitrite-induced methemoglobinemia in channel catfish (*Ictalurus punctatus*), *J. Fish. Res. Bd. Can.*, 36, 1141, 1979.
83. Huey, D. W., Simco, B. A., and Criswell, D. W., Nitrite-induced methemoglobin formation in channel catfish, *Trans. Am. Fish. Soc.*, 109, 558, 1980.
84. Colt, J., Ludwig, R., Tchobanoglous, G., and Cech, J. J., Jr., The effects of nitrite on the short-term growth and survival of channel catfish, *Ictalurus punctatus*, *Aquaculture*, 24, 111, 1981.
85. Tucker, C. S., Variability of percent methemoglobin in pond populations of nitrite-exposed channel catfish, *Prog. Fish-Cult.*, 45, 108, 1983.
86. Perry, W. G., Jr. and Avault, J. W., Jr., Culture of blue, channel and white catfish in brackish water ponds, *Proc. S.E. Assoc. Game Fish Comm.*, 23, 592, 1970.
87. Perry, W. G., Jr., Distribution and relative abundance of blue catfish, *Ictalurus punctatus*, with relation to salinity, *Proc. S.E. Assoc. Game Fish Comm.*, 21, 436, 1967.
88. Allen, K. O. and Avault, J. W., Jr., Effects of brackish water on ichthyophthiriasis of channel catfish, *Prog. Fish-Cult.*, 32, 227, 1970.

89. Johnson, S. K., Laboratory evaluation of several chemicals as preventatives of ich disease, in *Proc. 1976 Fish Farming Conf. Annu. Convention Catfish Farmers of Texas,* Texas A&M University, College Station, 1976, 91.
90. Lewis, S. D., Effect of selected concentrations of sodium chloride on the growth of channel catfish, *Proc. S.E. Assoc. Game Fish Comm.,* 25, 459, 1972.
91. Allen, K. O. and Avault, J. W., Jr., Effects of salinity on growth and survival of channel catfish, *Ictalurus punctatus, Proc. S.E. Assoc. Game and Fish Comm.,* 23, 319, 1970.
92. Allen, K. O. and Avault, J. W., Jr., Notes on the relative salinity tolerance of channel and blue catfish, *Prog. Fish-Cult.,* 33, 135, 1971.
93. Stickney, R. R. and Simco, B. A., Salinity tolerance of catfish hybrids, *Trans. Am. Fish. Soc.,* 100, 790, 1971.
94. Boyd, C. E., Preacher, J. W., and Justice, L., Hardness, alkalinity, pH, and pond fertilization, *Proc. S.E. Assoc. Fish Wildl. Agen.,* 32, 605, 1978.
95. McKee, J. E. and Wolf, H. W., Eds., Water Quality Criteria, 2nd ed., State Water Quality Control Board Publ. 3-A, State of California, Sacramento, 1963, 1.
96. Mandal, B. K. and Boyd, C. E., Reduction of pH in waters with high total alkalinity and low total hardness, *Prog. Fish-Cult.,* 42, 183, 1980.
97. Piper, R. G., McElwain, I. B., Orme, L. E., McCraren, J. P., Fowler, L. G., and Leonard, J. R., Fish Hatchery Management, Fish and Wildlife Service, U.S. Dpeartment of the Interior, Washington, D.C., 1982, 1.
98. Clemens, H. P. and Sneed, K. E., Spawning Behavior of the Channel Catfish *Ictalurus punctatus,* Special Scientific Report-Fisheries, No. 316, U.S. Department of the Interior, 1957, 1.
99. Brauhn, J. L., Fall spawning of channel catfish, *Prog. Fish-Cult.,* 33, 150, 1971.
100. Nelson, B., Spawning of channel catfish by use of hormone, *Proc. S.E. Assoc. Game Fish Comm.,* 14, 145, 1960.
101. Martin, M., Techniques of catfish fingerling production, in *Proc. 1967 Fish Farming Conf.,* Texas A&M University, College Station, 1967, 13.
102. Carter, R. R. and Thomas, A. E., Spawning channel catfish in tanks, *Prog. Fish-Cult.,* 39, 13, 1977.
103. Clemens, H. P. and Sneed, K. E., Bioassay and the Use of Pituitary Materials to Spawn Warmwater Fishes, Resources Report 61, U.S. Fish and Wildlife Service, Washington, D.C., 1962, 1.
104. Brooks, M. J., Smitherman, R. O., Chapell, J. A., and Dunham, R. A., Sex-weight relations in blue, channel, and white catfishes: implications for brood stock selection, *Prog. Fish-Cult.,* 44, 105, 1982.
105. Beaver, J. A., Sneed, K. E., and Dupree, H. K., The difference in growth of male and female channel catfish in hatchery ponds, *Prog. Fish-Cult.,* 28, 47, 1966.
106. El-Ibiary, H. M., Joyce, J. A., Page, J. W., and Hill, T. K., Comparison between sequential and concurrent matings of two females and one male channel catfish, *Ictalurus punctatus,* in spawning pens, *Aquaculture,* 10, 153, 1977.
107. Bondari, K., Efficiency of male reproduction in channel catfish, *Aquaculture,* 35, 79, 1983.
108. Clapp, A., Some experiments in rearing channel catfish, *Trans. Am. Fish. Soc.,* 59, 114, 1929.
109. Suppes, V. C., Jar incubation of channel catfish eggs, *Prog. Fish-Cult.,* 34, 48, 1972.
110. Toole, M., Channel catfish culture in Texas, *Prog., Fish-Cult.,* 13, 3, 1951.
111. Murphree, J. M., *Channel Culture Propagation,* Privately printed by T. J. Rennick.
112. Saksena, V. P., Yamamoto, K., and Riggs, C. D., Early development of the channel catfish, *Prog. Fish-Cult.,* 23, 156, 1961.
113. Reagan, R. E., Jr. and Conley, C. M., Effect of egg diameter on growth of channel catfish, *Prog. Fish-Cult.,* 39, 133, 1977.
114. Bryan, R. D. and Allen, K. O., Pond culture of channel catfish fingerlings, *Prog. Fish-Cult.,* 31, 38, 1969.
115. Bondari, K., Genetic and environmental control of fingerling size in channel catfish, *Aquaculture,* 34, 171, 1983.
116. El-Ibiary, H. M., Andrews, J. W., Joyce, J. A., Page, J. W., and DeLoach, H. L., Sources of variations in body size traits, dress-out weight, and lipid content and their correlations in channel catfish, *Ictalurus punctatus, Trans. Am. Fish. Soc.,* 105, 267, 1976.
117. Bondari, K., Response to bidirectional selection for body weight in channel catfish, *Aquaculture,* 33, 73, 1983.
118. Broussard, M. C., Jr. and Stickney, R. R., Growth of four strains of channel catfish in communal ponds, *Proc. S.E. Assoc. Fish Wildl. Agen.,* 35, in press.
119. Dunham, R. A. and Smitherman, R. O., Growth in response to winter feeding by blue, channel, white and hybrid catfishes, *Prog. Fish-Cult.,* 43, 63, 1981.
120. Giudice, J. J., Growth of a blue x channel catfish hybrid as compared to its parent species, *Prog. Fish-Cult.,* 28, 142, 1966.
121. Yant, D. R., Smitherman, R. O., and Green, O. L., Production of hybrid (blue x channel) catfish and channel catfish in ponds, *Proc. S.E. Assoc. Game Fish Comm.,* 29, 82, 1975.

122. Tave, D. and Smitherman, R. O., Spawning success of reciprocal hybrid pairings between blue and channel catfishes with an without hormone injection, *Prog. Fish-Cult.*, 44, 73, 1982.

123. Brooks, M. J., Smitherman, R. O., Chappell, J. A., and Dunham, R. A., Sex-weight relations in blue, channel, and white catfishes: implications for brook stock selection, *Prog. Fish-Cult.*, 44, 105, 1982.

124. Dunham, R. A., Smitherman, R. O., Brooks, M. J., Benchakan, M., and Chappell, J. A., Paternal predominance in reciprocal channel-blue hybrid catfish, *Aquaculture*, 29, 389, 1982.

125. Wolters, W. R., Libey, G. S., and Chrisman, C. L., Induction of triploidy in channel catfish, *Trans. Am. Fish. Soc.*, 110, 310, 1981.

126. Wolters, W. R., Libey, G. S., and Chrisman, C. L., Effect of triploidy on growth and gonad development of channel catfish, *Trans. Am. Fish. Soc.*, 111, 102, 1982.

127. Westerman, A. G. and Birge, W. J., Accelerated rate of albinism in channel catfish exposed to metals, *Prog. Fish-Cult.*, 40, 143, 1978.

128. Joint Subcommittee on Aquaculture, *National Aquaculture Development Plan*, Vol. 2, Washington, D.C., 1983, 17.

129. National Academy of Sciences, Nutrient Requirements of Trout, Salmon and Catfish, National Academy of Sciences, Washington, D.C., 1973, 1.

130. National Academy of Sciences, Nutrient Requirements of Warmwater Fishes, National Academy of Sciences, Washington, D.C., 1977, 1.

131. Stickney, R. R. and Lovell, R. T., Eds., Nutrition and Feeding of Channel Catfish, Southern Cooperative Series Bull. 218, 1977, 1.

132. Winfree, R. A., Starter Diets for Channel Catfish: Effects of Formulation on Growth and Body Composition, *Ph.D. dissertation, Texas A&M University, College Station*, 1983, 1.

133. Murai, T. and Andrews, J. W., Effects of frequency of feeding on growth and food conversion of channel catfish, *Bull. Jpn. Soc. Sci. Fish.*, 42, 159, 1976.

134. Andrews, J. W. and Page, J. W., The effects of frequency of feeding on culture of catfish, *Trans. Am. Fish. Soc.*, 104, 317, 1975.

135. Andrews, J. W., Some effects of feeding rate on growth, feed conversion and nutrient absorption of channel catfish, *Aquaculture*, 16, 243, 1979.

136. Greenland, D. C. and Gill, R. L., Multiple daily feedings with automatic feeders improve growth and feed conversion rates of channel catfish, *Prog. Fish-Cult.*, 41, 151, 1979.

137. Tucker, L., Boyd, C. E., and McCoy, E. W., Effects of feeding rate on water quality, production of channel catfish, and economic returns, *Trans. Am. Fish. Soc.*, 108, 389, 1979.

138. Bardach, J. E., Ryther, J. H., and McLarney, W. O., *Aquaculture*, Interscience, New York, 1972, 1.

139. Lovell, R. T. and Sirikul, B., Winter feeding of channel catfish, *Proc. S.E. Assoc. Game Fish Comm.*, 28, 208, 1974.

140. Randolph, K. N. and Clemens, H. P., Some factors influencing the feeding behavior of channel catfish in culture ponds, *Trans. Am. Fish. Soc.*, 105, 718, 1976.

141. Robinette, H. R., Busch, R. L., Newton, S. H., Haskins, C. J., Davis, S., and Stickney, R. R., Winter feeding of channel catfish in Mississippi, Arkansas, and Texas, *Proc. S.E. Assoc. Game Fish Comm.*, 36, 162, 1982.

142. Randolph, K. N. and Clemens, H. P., Home areas and swimways in channel catfish culture ponds, *Trans. Am. Fish. Soc.*, 105, 725, 1976.

143. Randolph, K. N. and Clemens, H. P., Effects of short term food deprivation on channel catfish and implications for culture practices, *Prog. Fish-Cult.*, 40, 48, 1978.

144. Robinette, H. R. and Dearing, A. S., Shrimp by-product meal in diets of channel catfish, *Prog. Fish-Cult.*, 40, 39, 1978.

145. Wilson, R. P. and Robinson, E. H., Protein and Amino Acid Nutrition for Channel Catfish, Information Bull. 25, Mississippi Agricultural and Forestry Experiment Station, Mississippi State, 1982, 1.

146. Lovell, R. T., Prather, E. E., Tres-Dick, J., and Chhorn, L., Effects of additional fish meal to all-plant feeds on the dietary protein needs of channel catfish in ponds, *Proc. S.E. Assoc. Game Fish Comm.*, 28, 222, 1974.

147. Dorsa, W. J., Robinette, H. R., Robinson, E. H., and Poe, W. E., Effects of dietary cottonseed meal and gossypol on growth of young channel catfish, *Trans. Am. Fish. Soc.*, 111, 651, 1982.

148. Liang, J. K. and Lovell, R. T., Biological evaluation of aquatic plants as potential ingredients in supplemental feeds for channel catfish, *Proc. S.E. Assoc. Game Fish Comm.*, 24, 638, 1971.

149. Robinette, H. R., Brunson, M. W., and Day, E. J., Use of duckweed in diets for channel catfish, *Proc. S.E. Assoc. Fish Wildl. Agen.*, 34, 108, 1980.

150. Liang, J. K. and Lovell, R. T., Nutritional value of water hyacinth in channel catfish feeds, *Hyacinth Cont. J.*, 9, 40, 1971.

151. Smith, B. W. and Lovell, R. T., Digestibility of nutrients in semi-purified rations by channel catfish in stainless steel troughs, *Proc. S.E. Assoc. Game Fish Comm.*, 25, 452, 1972.

152. Leary, D. F. and Lovell, R. T., Value of fiber in production-type diets for channel catfish, *Trans. Am. Fish. Soc.,* 104, 328, 1975.

153. Stickney, R. R. and Shumway, S. E., Occurrence of cellulase activity in the stomachs of fishes, *J. Fish Biol.,* 6, 779, 1974.

154. Stickney, R. R. and Andrews, J. W., Combined effects of dietary lipids and environmental temperature on growth, metabolism and body composition of channel catfish, *J. Nutr.,* 101, 1703, 1971.

155. Gatlin, D. M., III and Stickney, R. R., Fall-winter growth of young channel catfish in response to quantity and source of dietary lipid, *Trans. Am. Fish. Soc.,* 111, 90, 1982.

156. Stickney, R. R. and Andrews, J. W., Effects of dietary lipids on growth, food conversion, lipid and fatty acid composition of channel catfish, *J. Nutr.,* 102, 249, 1972.

157. Yingst, W. L., III and Stickney, R. R., Growth of caged channel catfish fingerlings reared on diets containing various lipids, *Prog. Fish-Cult.,* 42, 24, 1980.

158. Stickney, R. R., McGeachin, R. B., Lewis, D. H., and Marks, J., Response of young channel catfish to diets containing purified fatty acids, *Trans. Am. Fish. Soc.,* 112, 665, 1983.

159. White, A., Handler, P., and Smith, E. L., *Principles of Biochemistry,* McGraw-Hill, New York, 1964, 1.

160. Lovell, R. T. and Li, Y-P., Essentiality of vitamin D in diets of channel catfish (*Ictalurus punctatus),* *Trans. Am. Fish. Soc.,* 107, 809, 1978.

161. Andrews, J. W., Murai, T., and Page, J. W., Effects of dietary cholecalciferol and ergocalciferol on catfish, *Aquaculture,* 19, 49, 1980.

162. Murai, T. and Andrews, J. W., Vitamin K and anticoagulant relationships in catfish diets, *Bull. Jpn. Soc. Sci. Fish.,* 43, 785, 1977.

163. Lovell, R. T., Essentiality of vitamin C in feeds for intensively fed caged channel catfish, *J. Nutr.,* 103, 134, 1973.

164. Lovell, R. T. and Durve, V. S., Vitamin C and disease resistance in channel catfish, *Can. J. Fish. Aquat. Sci.,* 39, 948, 1982.

165. Andrews, J. W., Murai, T., and Campbell, C., Effects of dietary calcium and phosphorus on growth, food conversion, bone ash and hematocrit levels of catfish, *J. Nutr.,* 103, 766, 1973.

166. Lovell, R. T., Dietary phosphorus requirement of channel catfish (*Ictalurus punctatus),* *Trans. Am. Fish. Soc.,* 107, 617, 1978.

167. Murray, M. W. and Andrews, J. W., Channel catfish: the absence of an effect of dietary salt on growth, *Prog. Fish-Cult.,* 41, 155, 1979.

168. Murai, T., Andrews, J. W., and Smith, R. G., Effects of dietary copper on channel catfish, *Aquaculture,* 22, 353, 1981.

169. Meyer, F. P., The impact of diseases on fish farming, *Am. Fishes Trout News,* March-April 1967.

170. Plumb, J. A., Principle Diseases of Farm-Raised Catfish, Southern Cooperative Series Bull. No. 225, Auburn University, Auburn, Ala., 1979, 1.

171. Meyer, F. P., Schnick, R. A., Cumming, K. B., and Berger, B. L., Registration status of fishery chemicals, February, 1976, *Prog. Fish-Cult.,* 38, 3, 1976.

172. Meyer, F. P. and Bullock, G. L., *Edwardsiella tarda,* a new pathogen of channel catfish, *Ictalurus punctatus, Appl. Microbiol.,* 25, 155, 1973.

173. Hilton, L. R. and Wilson, J. L., Terramycin-resistant *Edwardsiella tarda* in channel catfish, *Prog. Fish-Cult.,* 42, 159, 1980.

174. Hawke, J. P., A bacterium associated with disease of pond cultured channel catfish, *Ictalurus punctatus, J. Fish. Res. Bd. Can.,* 36, 1508, 1979.

175. Hawke, J. P., McWhorter, A. C., Steigerwalt, A. G., and Brenner, D. J., *Edwardsiella ictaluri* sp. nov., the causative agent of enteric septicemia of catfish, *Int. J. Syst. Bact.,* 31, 396, 1981.

176. Plumb, J. A. and Sanchez, D. J., Susceptibility of five species of fish to *Edwardsiella ictaluri, J. Fish Dis.,* 6, 261, 1983.

177. Thune, R. L. and Plumb, J. A., Effect of delivery method and antigen preparation on the production of antibodies against *Aeromonas hydrophila* in channel catfish, *Prog. Fish-Cult.,* 44, 53, 1982.

178. Anon., For Fish Farmers, Newsletter from the Cooperative Extension Service, Mississippi State University, October 25, 1982.

179. Greenland, D. C., Recent developments in harvesting, grading, loading and hauling pond raised catfish, *Trans. Am. Soc. Agric. Eng.,* 17, 59, 1974.

180. Lovell, R. T. and Ammerman, G. R., Processing Farm-Raised Catfish, Southern Cooperative Series Bull. 193, Auburn University, Auburn, Ala., 1974, 1.

181. Thaysen, A. C., The origin of an earthy or muddy taint in fish. I. The nature and isolation of the taint, *Ann. Appl. Biol.,* 23, 99, 1936.

182. Thaysen, A. C. and Pentelow, F. T. K., The origin of an earthy or muddy taint in fish. II. The effect on fish of the taint produced by an odoriferous species of *Actinomyces, Ann. Appl. Biol.,* 23, 105, 1936.

183. Brown, S. W. and Boyd, C. E., Off-flavor in channel catfish from commercial ponds, *Trans. Am. Fish. Soc.,* 111, 379, 1982.

165

184. Lovell, R. T., New off-flavors in pond-cultured channel catfish, *Aquaculture,* 30, 329, 1983.
185. Tabachek, J. A. L. and Yurkowski, M., Isolation and identification of blue-green algae producing muddy odor metabolites, geosmin, and 2-methyliso-orneol, in saline lakes in Manitoba, *J. Fish. Res. Bd. Can.,* 33, 25, 1976.
186. Gerber, N. N. and Lechevalier, H. A., Geosmin, and earthy-smelling substance isolated from actinomycetes, *Appl. Microbiol.,* 13, 935, 1965.
187. Safferman, R. S., Rosen, A. A., Mashni, C. I., and Morris, M. E., Earthy-smelling substance from blue-green algae, *Environ. Sci. Technol.,* 1, 429, 1967.
188. Lovell, R. T. and Sackey, L. A., Absorption by channel catfish of earthy-musty flavor compounds by cultures of blue-green algae, *Trans. Am. Fish. Soc.,* 102, 774, 1973.
189. Yurkowski, M. and Tabachek, J. L., Identification, analysis, and removal of geosmin from muddy-flavored trout, *J. Fish. Res. Bd. Can.,* 31, 1851, 1974.
190. Iredale, D. G. and York, R. K., Purging a muddy-earthy flavor taint from rainbow trout (*Salmo gairdneri*) by transferring to artificial and natural holding environments, *J. Fish. Res. Bd. Can.,* 33, 160, 1976.
191. Cuenco, M. L., A Model of Fish Bioenergetics and Growth at the Organismal and Population Levels in Laboratory and Pond Environments, Ph.D. dissertation, Texas A&M University, College Station, 1982, 1.
192. Johnston, I. A., A comparative study of glycolysis in red and white muscles of the trout (*Salmo gairdneri*) and mirror carp (*Cyprinus carpio*), *J. Fish Biol.,* 11, 575, 1977.
193. Christiansen, B., Lomholt, J. P., and Johansen, K., Oxygen uptake of carp, *Cyrpinus carpio,* swimming in normoxic and hypoxic water, *Environ. Biol. Fish.,* 7, 291, 1982.
194. Balon, E. K., *Domestication of the Carp Cyrpinus carpio L.,* Royal Ontario Museum Life Sciences Misc. Publ., Toronto, Canada, 1974, 1.
195. Ling, S. W., *Aquaculture in Southeast Asia,* Washington Sea Grant Publication, College of Fisheries, University of Washington, Seattle, 1977, 103.
196. Bowen, J. R., A history of fish culture as related to the development of fisheries programs, in *A Century of Fisheries in North America,* Benson, N. G., Ed., American Fisheries Society, Washington, D. C., 1970, 71.
197. Doughty, R. W., Wildlife conservation in late nineteenth-century Texas: the carp experiment, *Southwest. Hist. Q.,* 84, 169, 1980.
198. George, T. T., The Chinese grass carp, *Ctenopharyngodon idella,* its biology, introduction, control of aquatic macrophytes and breeding in the Sudan, *Aquaculture,* 27, 317, 1982.
199. FAO, Yearbook of Fishery Statistics Catches and Landings, Food and Agriculture Organization, Rome, Italy, 1981, 1.
200. Krupauer, V., Pond Fish Culture in Czechoslovakia, EIFAC Occasional Paper No. 8, Food and Agriculture Organization, Rome, Italy, 1973, 1.
201. Hepher, B. and Pruginin, Y., *Commercial Fish Farming,* John Wiley & Sons, New York, 1981, 1.
202. Lovell, R. T., Fish Culture in Poland, International Center for Aquaculture, Research and Development Series No. 16, Agricultural Experiment Station, Auburn University, Auburn, Ala., 1977, 1.
203. Debeljak, L., Pleic, D., and Turk, M., Rearing of cyprinid fish fry and zooplankton in Croatia, in *Cultivation of Fish Fry and its Live Food,* Styczynska-Jurewicz, E., Backiel, T., Jaspers, E., and Personne, G., Eds., European Mariculture Society, Bredene, Belgium, 1979, 179.
204. Kafuku, T. and Ikenoue, H., *Modern methods of aquaculture in Japan,* Elsevier, New York, 1983, 1.
205. Tapidor, D. D., Henderson, H. F., Delmendo, M. N., and Tsutsui, H., Freshwater Fisheries and Aquaculture in China, FAO Fisheries Tech. Paper No. 168, FIR/T168, Food and Agriculture Organization, Rome, Italy, 1977, 1.
206. FAO, Aquaculture Development in China, Aquaculture Development and Coordination Program, ADCP/REP/79/10, Food and Agriculture Organization, Rome, Italy, 1979, 1.
207. Martin, J. M., personal communication, 1984.
208. Stevenson, J. M., Observations on grass carp in Arkansas, *Prog. Fish-Cult,* 27, 203, 1965.
209. Bailey, W. M., Meyer, F. P., Martin, J. M., and Gray, D. L., Fish farm production in Arkansas during 1972, *Proc. S.E. Assoc. Game Fish. Comm.,* 27, 750, 1973.
210. Henderson, S. and Freeze, M., The Aquaculture Industry of Arkansas in 1978, Arkansas Game and Fish Commission, Little Rock, 1979, 1.
211. Chervinski, J., personal communication, 1979.
212. Trzebiatowski, R., Rearing carp juveniles in heated waste-water of a power station, in *Cultivation of Fish Fry and its Live Food,* Styczynska-Jurewicz, E., Backiel, T., Jaspers, E., and Personne, G., Eds., European Mariculture Society, Bredene, Belgium, 1979, 507.
213. Chadhuri, H., Chakrabarty, R. D., Sen, P. R., Rao, N. G., and Jena, S., A new high fish production in India with record yields by composite fish culture in freshwater ponds, *Aquaculture,* 6, 343, 1975.

214. Sinha, V. R. P. and Vijaya, G., On the growth of grass carp, *Ctenopharyngodon idella* Val., in composite fish culture at Kalyani, West Bengal (India), *Aquaculture,* 5, 283, 1975.

215. Malacha, S. R., Buck, H. D., Baur, R. J., and Onizuka, D. R., Polyculture of the freshwater prawn, *Macrobrachium rosenbergii*, Chinese and common carps in ponds enriched with swine manure. I. Initial trials, *Aquaculture,* 25, 101, 1981.

216. Rappaport, U. and Sarig, S., The effect of population density of carp in monoculture under conditions of intensive growth, *Bamidgeh,* 31, 26, 1979.

217. Bieniarz, K., Epler, P., Thuy, L-N., and Kogut, E., Changes in the ovaries of adult carp, *Aquaculture,* 17, 45, 1979.

218. Rothbard, S., Induced reproduction in cultivated cyprinids — the common carp and the group of Chinese carps. II. The rearing of larvae and the primary nursing of fry, *Bamidgeh,* 34, 20, 1982.

219. Rothbard, S., Induced reproduction in cultivated cyprinids — the common carp and the group of Chinese carps. I. The technique of induction, spawning and hatching, *Bamidgeh,* 33, 103, 1981.

220. Jalbert, J., Breton, B., Brzuska, E., Fostier, A., and Wieniawski, J., A new tool for induced spawning: the use of 17 alpha-hydroxy-20 beta-dihydroprogesterone to spawn carp at low temperature, *Aquaculture,* 10, 353, 1977.

221. Epler, P., Bieniarz, K., and Marosz, E., The effect of low doses of carp hypophysal homogenate and some steroid hormones on carp (*Cyprinus carpio* L.) oocyte maturation in vitro, *Aquaculture,* 26, 245, 1982.

222. Rabelahatra, A., Improvement of techniques for carp fry production in the laboratory, *Aquaculture,* 27, 307, 1982.

223. Bisshai, H. M., Ishak, M. M., and Labib, W., Fecundity of the mirror carp *Cyrpinus carpio* L. at the Serow Fish Farm (Egypt), *Aquaculture,* 4, 257, 1974.

224. Tomita, M., Iwahashi, M., and Suzuki, R., Number of spawned eggs and ovarian eggs and egg diameter and percent eyed eggs with reference to the size of female carp, *Bull. Jpn. Soc. Sci. Fish.,* 46, 1077, 1980.

225. Hulata, G. and Rothbard, S., Cold storage of carp semen for short periods, *Aquaculture,* 16, 267, 1979.

226. Soin, S. G., Some features of the development of the carp, *Cyrpinus carpio*, under hatchery conditions, *Vopr. Ikhtioogii,* 17, 759, 1977.

227. Brody, T., Moav, R., Abramson, Z. V., Hulata, G., and Wohlfarth, G., Applications of electrophoretic genetic markers for fish breeding. II. Genetic variations within maternal half-sibs in carp, *Aquaculture,* 9, 351, 1976.

228. Brody, T., Storch, N., Kirsht, D., Hulata, G., Wohlfarth, G., and Moav, R., Application of electrophoretic genetic markers to fish breeding. III. Dialleel analysis of growth rate in carp, *Aquaculture,* 20, 371, 1980.

229. Brody, T., Wohlfarth, G., Hulata, G., and Moav, R., Application of electrophoretic genetic markers to fish breeding. IV. Assessment of breeding value to full-sib families, *Aquaculture,* 21, 175, 1981.

230. Hines, R. S., Wohlfarth, G., Moav, R., and Hulata, G., Genetic differences in susceptibility to two diseases among strains of the common carp, *Aquaculture,* 3, 187, 1974.

231. Hulata, G., Moav, R., and Wohlfarth, G., Genetic differences between the Chinese and European races of the common carp. III. Gonad abnormalities in hybrids, *J. Fish Biol.,* 16, 369, 1980.

232. Hulata, G., Moav, R., and Wohlfarth, G., Effects of crowding and availability of food on growth rate of fry in the European and Chinese races of the common carp, *J. Fish Biol.,* 20, 323, 1982.

233. Sin, A. W-C., Stock improvement of the common carp in Hong Kong through hybridization with the introduced Israeli race 'Dor-70', *Aquaculture,* 29, 299, 1982.

234. Suzuki, R., Yamaguchi, M., and Ishikawa, K., Differences in the growth rate of two races of the common carp at various water temperatures, *Bull. Freshw. Fish. Res. Lab., Tokyo,* 27, 21, 1977.

235. Suzuki, R. and Yamaguchi, M., Improvement of quality of the common carp by crossbreeding, *Bull. Jpn. Soc. Sci. Fish.,* 46, 1427, 1980.

236. Wohlfarth, G., Moav, R., Hulata, G., and Beiles, A., Genetic variations in seine escapability of the common carp, *Aquaculture,* 5, 375, 1975.

237. Wohlfarth, G., Lahman, M., Hulata, G., and Moav, R., The story of "Dor-70", a selected strain of the Israeli common carp, *Bamidgeh,* 32, 3, 1980.

238. Wohlfarth, G., Moav, R., and Hulata, G., A genotype-environment interaction for growth rate in the common carp growing in intensively manured ponds, *Aquaculture,* 33, 187, 1983.

239. Nagy, A., Rajki, K., Horvath, L., and Csanyi, V., Investigation on carp, *Cyrpinus carpio* L. gynogenesis, *J. Fish Biol.,* 13, 215, 1978.

240. Gervai, J., Peter, S., Nagy, A., Horvath, L., and Csanyi, V., Induced triploidy in carp, *Cyrpinus carpio* L., *J. Fish Biol.,* 17, 667, 1980.

241. Stanley, J. G., Martin, J. M., and Jones, J. B., Gynogenesis as a possible method for producing monosex grass carp (*Ctenopharyngodon idella*), *Prog. Fish Cult.,* 37, 25, 1975.

242. Stanley, J. and Jones, J. B., Morphology of androgenic and gynogenic grass carp, *Ctenopharyngodon idella* (Valenciennes), *J. Fish Biol.,* 9, 523, 1976.

243. Shelton, W., personal communication, 1984.
244. Beck, M. L. and Biggers, C. J., Chromosomal investigation of *Ctenopharyngodon idella* X *Aristichthys nobilis* hybrids, *Experimentia*, 38, 39, 1982.
245. Beck, M. L. and Biggers, C. J., Erythrocyte measurements of diploid and triploid *Ctenopharyngodon idella* X *Hypophthalmichthys nobilis* hybrids, *J. Fish Biol.*, 22, 497, 1983.
246. Beck, M. L. and Biggers, C. J., Ploidy of hybrids between grass carp and bighead carp determined by morphological analysis, *Trans. Am. Fish. Soc.*, 112, 808, 1983.
247. Williamson, J. and Smitherman, R. O., Food habits of hybrid buffalofish, tilapia, Israeli carp and channel catfish in polyculture, *Proc. S.E. Assoc. Game Fish. Comm.*, 29, 86, 1975.
248. Spataru, P. and Hepher, B., Common carp predating on tilapia fry in a high density polyculture fish pond system, *Bamidgeh*, 29, 1977.
249. Zur, O., The appearance of chironomid larvae in ponds containing common carp (*Cyprinus carpio*), *Bamidgeh*, 31, 105, 1979.
250. Zur, O., The importance of chironomid larvae as natural feed and as a biological indicator of soil condition in ponds containing common carp (*Cyprinus carpio*) and tilapia (*Sarotherodon aurea*), *Bamidgeh*, 32, 66, 1980.
251. Zur, O. and Sarig, S., Observation on the feeding of common carp (*Cyprinus carpio*) on chironomid larvae, *Bamidgeh*, 32, 25, 1980.
252. Tamas, G. and Horvath, L., Growth of cyprinids under optimal zooplankton conditions, *Bamidgeh*, 28, 50, 1976.
253. Tamas, G., Rearing of common carp fry and mass cultivation of its food organisms in ponds, in *Cultivation of Fish Fry and its Live Food*, Styczyunska-Jurewicz, E., Backiel, T., Jaspers, E., and Personne, G., Eds., European Mariculture Society, Bredene, Belgium, 1971, 281.
254. Bryant, P. L. and Matty, A. J., Optimization of *Artemia* feeding rate for carp larvae (*Cyprinus carpio* L.), *Aquaculture*, 21, 203, 1980.
255. Vankatesh, B. and Shetty, H. P. C., Studies on the growth rate of the grass carp *Ctenopharyngodon idella* (Valenciennes) fed on two aquatic weeds and a terrestrial grass, *Aquaculture*, 13, 45, 1978.
256. Grygierek, E., The influence of phytophagous fish on pond zooplankton, *Aquaculture*, 2, 197, 1973.
257. Singh, S. B., Dey, R. K., and Reddy, P. V. G. K., Observations on feeding of young grass carp on mosquito larvea, *Aquaculture*, 12, 361, 1977.
258. Watkins, C. E., Shireman, J. V., and Rottman, R. W., Food habits of fingerling grass carp, *Prog. Fish-Cult.*, 43, 95, 1981.
259. DeSilva, S. S. and Weerakoon, D. E. M., Growth, food intake and evacuation rates of grass carp, *Ctenopharyngodon idella*, fry, *Aquaculture*, 25, 67, 1981.
260. Barash, H., Plavnik, I., and Moav, R., Integration of duck and fish farming: experimental results, *Aquaculture*, 27, 129, 1982.
261. Dimitrov, M., Mineral fertilization of carp ponds in polycultural rearing, *Aquaculture*, 3, 273, 1974.
262. Klekot, L., Pilot-scale field experiment on the utilization of biologically purified municipal-industrial waste waters for fish farming, in *Cultivation of Fish Fry and its Live Food*, Styczynska-Jurewicz, E., Backiel, T., Jaspers, E., and Personne, G., Eds., European Mariculture Soc., Bredene, Belgium, 1979, 475.
263. Szlauer, L. and Szlauer, B., Utilization of waste effluent of a fertilizer plant for rearing of zooplankton and carp fry, in *Cultivation of Fish Fry and its Live Food*, Styczynska-Jurewicz, E., Backiel, T., Jaspers, E., and Persoone, G., Eds., European Mariculture Soc., Bredene, Belgium, 1979, 501.
264. Viola, S. and Arieli, Y., Evaluation of different grains as basic ingredients in complete diets for carp and tilapia in intensive culture, *Bamidgeh*, 35, 38, 1983.
265. Viola, S., Arieli, Y., Rappaport, U., and Mokady, S., Experiments in the nutrition of carp: replacement of fishmeal by soybeanmeal, *Bamidgeh*, 33, 35, 1981.
266. Viola, S., Mokady, S., Rappaport, U., and Arieli, Y., Partial and complete replacement of fish meal by soybean meal in feed for intensive culture of carp, *Aquaculture*, 26, 223, 1982.
267. Viola, S. and Rappaport, U., The "extra-caloric effect" of oil in the nutrition of carp, *Bamidgeh*, 31, 51, 1979.
268. Viola, S., Rappaport, U., Arieli, Y., Amidan, G., and Mokady, S., The effects of oil-coated pellets on carp (*Cyprinus carpio*) in intensive culture, *Aquaculture*, 26, 49, 1981.
269. Atack, T. H., Jauncey, K., and Matty, A. J., The utilization of some single cell proteins by fingerling mirror carp (*Cyprinus carpio*), *Aquaculture*, 18, 337, 1979.
270. Anwar, A., Ishak, M. M., El-Zeiny, M., and Hassanan, G. D., Activated sewage sludge as a replacement for bran-cotton seed meal mixture for carp, *Cyprinus carpio* L., *Aquaculture*, 28, 321, 1982.
271. Capper, B. S., Wood, J. F., and Jackson, A. J., The feeding value for carp of two types of mustard seed cake from Nepal, *Aquaculture*, 29, 373, 1982.
272. Christensen, M. S., Preliminary tests on the suitability of coffee pulp in the diets of common carp (*Cyprinus carpio* L.) and catfish (*Clarias mossambicus* Peters), *Aquaculture*, 25, 235, 1981.
273. Ufodike, E. B. C. and Matty, A. J., Growth responses and nutrient digestibility in mirror carp (*Cyprinus carpio*) fed different levels of cassava and rice, *Aquaculture*, 31, 41, 1983.

274. Lone, K. P. and Matty, A. J., Uptake and disappearance of radioactivity in blood and tissue of carp (*Cyprinus carpio*) after feeding 3 H-testosterone, *Aquaculture,* 24, 315, 1981.
275. Lone, K. P. and Matty, A. J., The effect of ethylestrenol on growth, food conversion and tissue chemistry of the carp, *Cyprinus carpio, Aquaculture,* 32, 39, 1983.
276. Lone, K. P. and Matty, A. J., Cellular effects of adrenosterone feeding to juvenile carp, *Cyprinus carpio* L., effect on liver, kidney, brain and muscle protein and nucleic acids, *J. Fish Biol.,* 21, 33, 1982.
277. Matty, A. J. and Lone, K. P., The effect of androgenic steroids as dietary additives on the growth of carp (*Cyprinus carpio*), *Proc. World Maricult. Soc.,* 10, 735, 1979.
278. Appelbaum, S. and Dor, U., Ten day experimental nursing of carp (*Cyprinus carpio* L.) larvae with dry feed, *Bamidgeh,* 30, 85, 1978.
279. Bryant, P. L. and Matty, A. J., Adaptation of carp (*Cyprinus carpio*) larvae to artificial diets. I. Optimum feeding rate and adaptation age for a commercial diet, *Aquaculture,* 23, 275, 1981.
280. Dabrowski, K., Dabrowski, H., and Grundniewski, C., A study of the feeding of common carp larvae with artificial food, *Aquaculture,* 13, 257, 1978.
281. Goolish, E. M. and Adelman, I. R., Effects of ration size and temperature on the growth of juvenile common carp (*Cyprinus carpio* L.), *Aquaculture,* 36, 27, 1984.
282. Dabrowski, K. and Kozak, B., The use of fish meal and soybean meal as a protein source in the diet of grass carp fry, *Aquaculture,* 18, 107, 1979.
283. Huisman, E. A. and Valentijn, P., Conversion efficiencies in grass carp (*Ctenopharyngodon idella,* Val.) using a food for commercial production, *Aquaculture,* 22, 279, 1981.
284. Dabrowski, K., Comparative aspects of protein digestion and amino acid absorption in fish and other animals, *Comp. Biochem. Physiol.,* 74A, 417, 1983.
285. Dabrowski, H., Grudniewski, C., and Dabrowski, K., Artificial diets for the common carp: effect of the addition of enzyme extracts, *Prog. Fish-Cult.,* 41, 196, 1979.
286. Farkas, T., Csengeri, I., Majoros, F., and Olah, J., Metabolism of fatty acids in fish. I. Development of essential fatty acid deficiency in the carp, *Cyprinus carpio* Linnaeus 1758, *Aquaculture,* 11, 147, 1977.
287. Farkas, T., Scengeri, I., Majoros, F., and Olah, J., Metabolism of fatty acids in fish. II. Biosynthesis of fatty acids in relation to diet in the carp, *Cyprinus carpio* Linnaeus 1758, *Aquaculture,* 14, 57, 1978.
288. Farkas, Csengeri, I., Majoros, F., and Olah, J., Metabolism of fatty acids in fish. III. Combined effect of environmental temperature and diet formulation and deposition of fatty acids in the carp, *Cyprinus carpio* Linnaeus 1758, *Aquaculture,* 20, 29, 1980.
289. Jonas, E., Ragyanszki, M., Olah, J., and Boross, L., Proteolytic digestive enzymes of carnivorous (*Silurus glanis* l.), herbivorous (*Hypophthalmichthys molitrix* Val.) and omnivorous (*Cyprinus carpio* L.) fishes, *Aquaculture,* 30, 145, 1983.
290. Ogino, C. and Saito, K., Protein nutrition in fish. I. The utilization of dietary protein by yound carp, *Bull. Jpn. Soc. Sci. Fish.,* 36, 250, 1970.
291. Ogino, C., Chiou, J. Y., and Takeuchi, T., Protein nutrition in fish. IV. Effects of dietary energy sources on the utilization of proteins by rainbow trout and carp, *Bull. Jpn. Soc. Sci. Fish.,* 42, 213, 1976.
292. Plakas, S. M., Katayama, T., Tanaka, Y., and Deshimaru, O., Changes in the level of circulating plasma free amino acids of carp (*Cyprinus carpio*) after feeding a protein and an amino acid diet of similar composition, *Aquaculture,* 21, 307, 1980.
293. Plakas, S. M. and Katayama, T., Apparent digestibilities of amino acids from three regions of the gastrointestinal tract of carp (*Cyprinus carpio*) after ingestion of a protein and a corresponding free amino acid diet, *Aquaculture,* 24, 309, 1981.
294. Sen, P. R., Rao, N. G. S., Gosh, S. R., and Rout, M., Observations on the protein and carbohydrate requirements of carps, *Aquaculture,* 13, 245, 1978.
295. Steffens, W., Protein utilization by rainbow trout (*Salmo gairdneri)* and carp (*Cyprinus carpio*): a brief review, *Aquaculture,* 23, 337, 1981.
296. Viola, S., Energy values of feedstuffs for carp, *Bamidgeh,* 29, 29, 1977.
297. Viola, S. and Amidan, G., The effects of different dietary oil supplements on the composition of carp's body fat, *Bamidgeh,* 30, 104, 1978.
298. Viola, S. and Arieli, Y., Nutrition studies with a high-protein pellet for carp and *Sarotherodon* spp. (tilapia), *Bamidgeh,* 34, 39, 1982.
299. Zeitler, M., Kirchghessner, M., and Schwarz, F. J., Effects of different protein and energy supplies on carcass composition of carp (*Cyprinus carpio* L.), *Aquaculture,* 36, 37, 1984.
300. Murata, H. and Higashi, T., Selective utilization of fatty acid as energy source in carp, *Bull. Jpn. Soc. Sci. Fish.,* 46, 1333, 1980.
301. Ogino, C., Protein requirements of carp and rainbow trout, *Bull. Jpn. Soc. Sci. Fish.,* 46, 385, 1980.
302. Ogino, C. and Chiou, J. Y., Mineral requirements in fish. II. Magnesium requirements in carp, *Bull. Jpn. Soc. Sci. Fish.,* 42, 71, 1976.

303. Ogino, C. and Kamizono, M., Mineral requirements in fish. I. Effects of dietary salt-mixture levels on growth, mortality, and body composition in rainbow trout and carp, *Bull. Jpn. Soc. Sci. Fish.*, 41, 429, 1975.

304. Ogino, C. and Takeda, H., Mineral requirements in fish. III. Calcium and phosphorus requirements in carp, *Bull. Jpn. Soc. Sci. Fish.*, 42, 793, 1976.

305. Ogino, C., Takeuchi, L., Takeda, H., and Watanabi, T., Availability of dietary phosphorus in carp and rainbow trout, *Bull. Jpn. Soc. Sci. Fish.*, 45, 1527, 1979.

306. Ogino, C., Watanabi, T., Kakino, J., Iwanaga, N., and Mizunox, M., B vitamin requirements of carp. III. Requirement for biotin, *Bull. Jpn. Soc. Sci. Fish.*, 36, 734, 1970.

307. Ogino, C., Cowey, C. B., and Chiou, J-Y., Leaf protein concentrate as a protein source in diets for carp and rainbow trout, *Bull. Jpn. Soc. Sci. Fish.*, 44, 49, 1978.

308. Ogino, C. and Yang, G-Y., Requirement of carp for dietary zinc, *Bull. Jpn. Soc. Sci. Fish.*, 45, 967, 1979.

309. Ogino, C. and Yang, G-Y., Requirements of carp and rainbow trout for dietary manganese and copper, *Bull. Jpn. Soc. Sci. Fish.*, 46, 455, 1980.

310. Sakamoto, S. and Yone, Y., Iron deficiency symptoms of carp, *Bull. Jpn. Soc. Sci. Fish.*, 44, 1157, 1978.

311. Sato, M., Yoshinaka, R., Yamamoto, Y., and Ikeda, S., Nonessentiality of ascorbic acid in the diet of carp, *Bull. Jpn. Soc. Sci. Fish.*, 44, 1151, 1978.

312. Shimeno, S., Takeda, M., Takayama, S., Fukui, A., Sasaki, H., and Kajiyama, H., Adaptation of hepatopancreatic enzymes to dietary carbohydrate in carp, *Bull. Jpn. Soc. Sci. Fish.*, 47, 71, 1981.

313. Takeuchi, T. and Watanabi, T., Requirement of carp for essential fatty acids, *Bull. Jpn. Soc. Sci. Fish.*, 43, 541, 1977.

314. Takeuchi, T., Watanabi, T., and Ogino, C., Use of hydrogenated fish oil and beef tallow as a dietary source for carp and rainbow trout, *Bull. Jpn. Soc. Sci. Fish.*, 44, 875, 1978.

315. Takeuchi, T., Watanabi, T., and Ogino, C., Digestibility of hydrogenated fish oils in carp and rainbow trout, *Bull. Jpn. Soc. Sci. Fish.*, 45, 1521, 1979.

316. Takeuchi, T., Watanabi, T., and Ogino, C., Availability of carbohydrate and lipid as dietary energy sources for carp, *Bull. Jpn. Soc. Sci. Fish.*, 45, 977, 1979.

317. Takeuchi, T., Watanabi, T., and Ogino, C., Optimum ratio of dietary energy to protein for carp, *Bull. Jpn. Soc. Sci. Fish.*, 45, 983, 1979.

318. Takeuchi, L., Takeuchi, T., and Ogino, C., Riboflavin requirement in carp and rainbow trout, *Bull. Jpn. Soc. Sci. Fish.*, 46, 733, 1980.

319. Watanabi, T., Takashima, F., Ogino, C., and Hibiya, T., Requirement of young carp for alpha-tocopherol, *Bull. Jpn. Soc. Sci. Fish.*, 36, 972, 1970.

320. Watanabi, T., Takeuchi, T., and Ogino, C., Effect of dietary methyl lineolate and linolenate on growth of carp, *Bull. Jpn. Soc. Sci. Fish.*, 41, 263, 1975.

321. Watanabi, T., Utsue, O., Kobayshi, I., and Ogino, C., Effect of dietary methyl linoleate and linolenate on growth of carp, *Bull. Jpn. Soc. Sci. Fish.*, 41, 257, 1975.

322. Yone, Y., and Toshima, N., The utilization of phosphorus in fish meal by carp and black sea bream, *Bull. Jpn. Soc. Sci. Fish.*, 45, 753, 1979.

323. Jauncey, K., Carp (*Cyprinus carpio* L.) nutrition — a review, in *Recent Advances in Aquaculture*, Muri, J. F. and Roberts, J. R., Eds., Westview Press, Boulder, Colo., 1982, 215.

324. Amlacher, E., *Textbook of Fish Diseases*, Conroy, D. A. and Herman, R. L., transl., T. F. H. Publications, Hong Kong, 1970, 1.

325. Bootsma, R. and Clerx, J. P. M., Columnaris disease of cultured carp *Cyprinus carpio* L., Characterization of the causative agent, *Aquaculture*, 7, 317, 1976.

326. Brady, L. and Hulsey, A., Propagation of buffalo fishes, *Proc. S.E. Assoc. Game Fish Comm.*, 13, 80, 1959.

327. Walter, M. C. and Frank, P. T., The propagation of buffalo, *Prog. Fish-Cult.*, 14, 129, 1952.

328. Swingle, H. S., Revised procedures for commercial production of bigmouth buffalo in ponds in the southeast, *Proc. S.E. Assoc. Game Fish Comm.*, 10, 162, 1956.

329. Tuten, J. S. and Avault, J. W., Growing red swamp crayfish (*Procambarus clarkii*) and several North American fish species together, *Prog. Fish-Cult.*, 43, 97, 1981.

330. Pullin, R. S. V., Choice of tilapia species for aquaculture (Abstr.), in *Int. Symp. Tilapia in Aquaculture*, Nazareth, Israel, May 8 to 13, 1983, 8.

331. Boulenger, G. A., Catalogue of the freshwater fishes of Africa, III, *Br. Mus. Nat. Hist. Lond.*, 1, 1915.

332. Copley, H., The tilapia of Kenya colony, *East Afr. Agricult. J.*, 18, 30, 1952.

333. Atz, J. W., The peregrinating tilapia, *Anim. King.*, 57, 148, 1954.

334. Chimits, P., Tilapia and its culture, *FAO Fish. Bull.*, 8, 1, 1955.

335. Chimits, P., The tilapias and their culture, A second review and bibliography, *FAO Fish. Bull.*, 10, 1, 1957.

336. Philippart, J-Cl. and Ruwet, J-Cl., Ecology and distribution of tilapias, in *The Biology and Culture of Tilapia*, Pullin, R. S. V. and Lowe-McConnell, R. H., Eds., International Center for Living Aquatic Resources Management, Manila, Philippines, 1982, 15.
337. Balarin, J. D. and Hatton, J. P., *Tilapia — a Guide to their Biology and Culture in Africa*, University of Stirling, 1979, 1.
338. Chervinski, J., Environmental physiology of tilapias, in, *The Biology and Culture of Tilapia*, Pullin, R. S. V. and Lowe-McConnell, R. H., Eds., International Center for Living Aquatic Resources Management, Manila, Philippines, 1982, 119.
339. Hepher, B. and Pruginin, Y., Tilapia culture in ponds under controlled conditions, in *The Biology and Culture of Tilapia*, Pullin, R. S. V. and Lowe-McConnell, R. H., Eds., International Center for Living Aquatic Resources Management, Manila, Philippines, 1982, 185.
340. Coche, A. G., Cage culture of tilapias, in *The Biology and Culture of Tilapia*, Pullin, R. S. V. and Lowe-McConnell, R. H., Eds., International Center for Living Aquatic Resources Management, Manila, Philippines, 1982, 205.
341. Schoenen, P., *A Bibliography of Important Tilapias (Pisces:Cichlidae) for Aquaculture*, Bibliographies 3, International Center for Living Aquatic Resources Management, Manila, Philippines, 1982, 1.
342. Payne, A. I. and Collinson, R. I., A comparison of the biological characteristics of *Sarotherodon niloticus* (L.) with those of *S. aureus* (Steindachner) and other tilapia of the delta and lower Nile, *Aquaculture*, 30, 335, 1983.
343. Avtalion, R. R., Duczyminer, M., Wojdani, A., and Pruginin, Y., Determination of allogenic and xenogenic markers in the genus *Tilapia*. II. Identification of *T. aurea, T. vulcani* and *T. nilotica* by electrophoretic analysis of their serum proteins, *Aquaculture*, 7, 255, 1976.
344. Chung, K. S., Effects of temperature on growth, survival, acclimation rate and body temperature in *Oreochromis (Tilapia) mossambicus*, (Abstr.), in Int. Symp. Tilapia in Aquaculture, Nazareth, Israel, May 8 to 13, 1983, 30.
345. Platt, S. and Hauser, W. J., Optimum temperature for feeding and growth of *Tilapia zillii*, *Prog. Fish-Cult.*, 40, 105, 1978.
346. Allanson, B. R. and Noble, R. G., The tolerance of *Tilapia mossambica* (Peters) to high temperature, *Trans. Am. Fish. Soc.*, 93, 323, 1964.
347. Denzer, H. W., Studies on the physiology of young *Tilapia*, *FAO Fish. Rep.*, 44, 356, 1968.
348. Gleastine, B. W., A study of the Cichlid *Tilapia aurea* (Steindachner) in a Thermally Modified Texas reservoir, M.S. thesis, Texas A&M University, College Station, 1974, 1.
349. McBay, L. G., The biology of *Tilapia nilotica* Linnaeus, *Proc. S.E. Assoc. Game Fish Comm.*, 15, 208, 1961.
350. Sarig, S., Winter storage of *Tilapia*, *FAO Fish Cult. Bull.*, 2, 8, 1969.
351. Shafland, P. L. and Pestrak, J. M., Lower lethal temperatures for fourteen nonnative fishes in Florida, *Environ. Biol. Fish.*, 7, 149, 1982.
352. Kelly, H. D., Preliminary studies on *Tilapia mossambica* Peters relative to experimental pond culture, *Proc. S.E. Assoc. Game Fish Comm.*, 10, 139, 1956.
353. Chervinski, J. and Stickney, R. R., Overwintering facilities for tilapia in Texas, *Prog. Fish-Cult.*, 43, 20, 1981.
354. Stickney, R. R. and Winfree, R. A., Tilapia overwintering systems, in *Proc. 1982 Fish Farming Conf. Annu. Convention Fish Farmers of Texas*, 1982, 58.
355. Henderson-Arzapalo, A., Design and Operational Characteristics of a Tilapia Rearing System for Temperate Regions, Ph.D. dissertation, Texas A&M University, College Station, 1984, 1.
356. Chervinski, J. and Yashouv, A., Preliminary experiments on the growth of *Tilapia aurea* (Steindachner) (Pisces, Cichlidae) in sea water ponds, *Bamidgeh*, 23, 125, 1971.
357. Rahimaldin, S. A., The Effect of Salinity on the Growth of Blue Tilapia (*Tilapia aurea*), M.S. thesis, Texas A&M University, College Station, 1983, 1.
358. Chervinski, J. and Zorn, M., Note on the growth of *Tilapia aurea* (Steindachner) and *Tilapia zillii* (Gervais) in sea-water ponds, *Aquaculture*, 4, 249, 1974.
359. Fishelson, L. and Popper, D., Experiments on rearing fish in salt water near the Dead Sea, Israel, *FAO Fish. Rep.*, 44, 244, 1968.
360. Loya, L. and Fishelson, L., Ecology of fish breeding in brackish water ponds near the Dead Sea (Israel), *J. Fish Biol.*, 1, 261, 1969.
361. Popper, D. and Lichatowich, T., Preliminary success in predator control of *Tilapia mossambica*, *Aquaculture*, 5, 213, 1975.
362. Potts, W. T. W., Foster, M. A., Rudy, P. O., and Howells, G. P., Sodium and water balance in the cichlid teleost, *Tilapia mossambica*, *J. Exp. Biol.*, 47, 461, 1967.
363. Maruyama, T., An observation on *Tilapia mossambica* in ponds referring to the diurnal movement with temperature change, *Bull. Freshw. Fish. Res. Lab. Tokyo*, 8, 25, 1958.

364. Magid, A. and Babiker, M. M., Oxygen consumption and respiratory behaviour of three Nile fishes, *Hydrobiology,* 46, 359, 1975.
365. Lovshin, L. L., da Silva, A. B., and Fernandes, J. A., The intensive culture of all male hybrids of *Tilapia hornorum (male) and T. nilotica* (female) in northeast Brazil, *FAO Fish. Rep.* 159, 162, 1977.
366. Stickney, R. R., Rowland, L. O., and Hesby, J. H., Water quality- *Tilapia aurea* interactions in ponds receiving swine and poultry wastes, *Proc. World Maricult. Soc.,* 8, 55, 1977.
367. Redner, B. D. and Stickney, R. R., Acclimation to ammonia by *Tilapia aurea, Trans. Am. Fish. Soc.,* 108, 383, 1979.
368. Almazan, G. and Boyd, C. E., Plankton production and tilapia yield in ponds, *Aquaculture,* 15, 75, 1978.
369. Sarig, S. and Marek, M., Results of intensive and semi-intensive fish breeding techniques in Israel in 1971—1973, *Bamidgeh,* 26, 28, 1974.
370. Suffern, J. S., Adams, S. M., Blaylock, B. G., Coutant, C. C., and Guthrie, C. A., Growth of monosex hybrid tilapia in the laboratory and sewage oxidation ponds, in *Symp. Culture of Exotic Fishes,* Fish Culture Section, American Fisheries Society, Auburn, Ala., 65, 1978.
371. Henderson-Arzapalo, A. and Stickney, R. R., Effects of stocking density on two tilapia species raised in an intensive culture system, *Proc. S.E. Assoc. Fish Wildl. Agen.,* 34, 379, 1980.
372. Henderson-Arzapalo, A., Stickney, R. R., and Lewis, D. H., Immune hypersensitivity in intensively cultured tilapia species, *Trans. Am. Fish. Soc.,* 109, 244, 1980.
373. Yashouv, A., Mixed fish culture in ponds and the role of tilapia in it, *Bamidgeh,* 21, 75, 1969.
374. Brick, R. W. and Stickney, R. R., Polyculture of *Tilapia aurea* and *Macrobrachium rosenbergii* in Texas, *Proc. World Maricult. Soc.,* 10, 222, 1979.
375. Rouse, D. B., The Evaluation of *Macrobrachium rosenbergii* and *Tilapia aurea* Polyculture in Texas, Ph.D. dissertation, Texas A&M University, College Station, 1981, 1.
376. Wilson, J. L. and Hilton, L. R., Effects of tilapia densities on growth of channel catfish, *Prog. Fish-Cult.,* 44, 207, 1982.
377. Boyd, C. E., Nitrogen fertilizer effects on production of *Tilapia* in ponds fertilized with phosphorus and potassium, *Aquaculture,* 7, 385, 1976.
378. Schroeder, G., Use of cowshed manure in fish ponds, *Bamidgeh,* 26, 84, 1974.
379. Edwards, P., A review of recycling organic wastes in fish, with emphasis on the tropics, *Aquaculture,* 21, 261, 1980.
380. Edwards, P. and Sinchumpasak, O-A., The harvest of microalgae from the effluent of a sewage fed high rate stabilization pond by *Tilapia nilotica.* I. Description of the system and the study of the high pond, *Aquaculture,* 23, 83, 1981.
381. Edwards, P., Sinchumpasak, O-A., and Tabucanon, M., The harvest of microalgae from the effluent of a sewage fed high rate stabilization pond by *Tilapia nilotica.* II. Studies of the fish ponds, *Aquaculture,* 23, 107, 1981.
382. Edwards, P., Sinchumpasak, O-A., Labhsetwar, V. K., and Tabucanon, M., The harvest of microalgae from the effluent of a sewage fed high rate stabilization pond by *Tilapia nilotica.* III. Maize cultivation experiment, bacteriological studies, and economic assessment, *Aquaculture,* 23, 149, 1981.
383. Stickney, R. R. and Hesby, J. H., Tilapia production in ponds receiving swine wastes, in *Symp. Culture of Exotic Fishes,* Fish Culture Section, American Fisheries Society, Auburn, Ala., 1978, 90.
384. Stickney, R. R., Hesby, J. H., McGeachin, R. B., and Isbell, W. A., Growth of *Tilapia nilotica* in ponds with differing histories of organic fertilization, *Aquaculture,* 17, 189, 1979.
385. Burns, R. P. and Stickney, R. R., Growth of *Tilapia aurea* in ponds receiving poultry wastes, *Aquaculture,* 20, 117, 1980.
386. McGeachin, R. B. and Stickney, R. R., Manuring rates for production of blue tilapia in simulated sewage lagoons receiving laying hen waste, *Prog. Fish-Cult.,* 44, 25, 1982.
387. Aguilar, V. L., Feasibility of growing tilapia fish in swine waste lagoons (Abstr.), in Int. Symp. on Tilapia in Aquaculture, Nazareth, Israel, May 8 to 13, 1983, 78.
388. Broussard, M. C., personal communication, 1983.
389. Spataru, P., Natural feed of *Tilapia aurea* Steindachner in polyculture, with supplementary feed and intensive manuring, *Bamidgeh,* 28, 57, 1976.
390. Doroshev, S. I. and Cornacchia, J. W., Initial swim bladder inflation in the larvae of *Tilapia mossambica* (Peters) and *Morone saxatilis* (Walbaum), *Aquaculture,* 16, 57, 1979.
391. Blay, J., Jr., Fecundity and spawning frequency of *Sarotherodon galilaeus* in a concrete pond, *Aquaculture,* 25, 95, 1981.
392. Babiker, M. M. and Ibrahim, H., Studies on the biology of reproduction in the cichlid *Tilapia nilotica* (L.): gonadal maturation and fecundity, *J. Fish. Biol.,* 14, 437, 1979.
393. Uchida, R. N. and King, J. E., Tank culture of tilapia, *Fish. Bull.,* 62, 21, 1962.
394. Huet, M., *Traite de Pisciculture,* Editions Ch. de Wyngaert, Bussels, 1970, 1.
395. Silvera, P. A. W., Factors Affecting Fry Production in *Sarotherodon niloticus* (Linnaeus), M.S. thesis, Auburn University, Auburn, Ala., 1978, 1.

396. Pierce, B. A., Production of hybrid tilapia in indoor aquaria, *Prog. Fish-Cult.*, 42, 233, 1980.
397. Berrios-Hernandez, J. M., Comparison of methods for reducing fry losses to cannibalism in tilapia production, *Prog. Fish-Cult.*, 45, 116, 1983.
398. Avault, J. W. and Shell, W. E., Preliminary studies with hybrid *Tilapia nilotica* and *T. mossambica*. *FAO Fish. Rep.*, 44, 237, 1967.
399. Rothbard, S. and Pruginin, Y., Induced spawning and artificial incubation of *Tilapia*, *Aquaculture*, 5, 315, 1975.
400. Rothbard, S., Observations on the reproductive behavior of *Tilapia zillii* and several *Sarotherodon* spp. under aquarium conditions, *Bamidgeh*, 31, 35, 1979.
401. Rothbard, S. and Hulata, G., Closed-system incubator for cichlid eggs, *Prog. Fish-Cult.*, 42, 203, 1980.
402. Tave, D. and Smitherman, R. O., Predicted response to selection for early growth in *Tilapia nilotica*, *Trans. Am. Fish. Soc.*, 109, 439, 1980.
403. Noble, R. L., Germany, R. D., and Hall, C. R., Interactions of blue tilapia and largemouth bass in a power plant cooling reservoir, *Proc. S.E. Assoc. Game Fish Comm.*, 29, 247, 1975.
404. Guerrero, R. D., Control of tilapia reproduction, in *The Biology and Culture of Tilapia*, Pullin, R. S. V. and Lowe-McConnell, R. H., Eds., International Center for Living Aquatic Resources Management, Manila, Philippines, 1982, 309.
405. Lowe-McConnell, R. H., Observations on the biology of *Tilapia nilotica* Linné in east Africa waters, *Rev. Zool. Bot. Afr.*, 57, 131, 1958.
406. Van Someren, V. D. and Whitehead, P. J., The culture of *Tilapia nigra* (Gunther) in ponds. III. The early growth of males and females at comparable stocking rates, and the length/weight relationship, *East Afr. Agric. For. J.*, 25, 169, 1960.
407. Van Someren, V. D. and Whitehead, P. J., The culture of *Tilapia nigra* (Gunther) in ponds. IV. The seasonal growth of male *T. nigra*, *East Afr. Agric. For. J.*, 26, 79, 1960.
408. Fryer, G. and Iles, T. D., *The Cichlid Fishes of the Great Lakes of Africa: Their Biology and Evolution*, T. F. H. Publishers, Neptune City, N.J., 1972, 1.
409. Guerrero, R. D. and Guerrero, L. A., Monosex culture of male and female *T. mossambica* in ponds at three stocking rates, *Kalikasan: Philipp, J. Biol.*, 4, 129, 1975.
410. Eyeson, K. N., Stunting and reproduction in pond-reared *Sarotherodon melanotheron*, *Aquaculture*, 31, 257, 1983.
411. Pagan, F. A., Cage culture of tilapia, *FAO Fish Cult. Bull.*, 2, 6, 1969.
412. Pagan-Font, F. A., Cage culture as a mechanical method for controlling reproduction of *Tilapia aurea* (Steindachner), *Aquaculture*, 6, 243, 1975.
413. Rifai, S. A., Control of reproduction of *Tilapia nilotica* using cage culture, *Aquaculture*, 20, 177, 1980.
414. Dunseth, D. R. and Bayne, D. R., Recruitment control and production of *Tilapia aurea* (Steindachner) with the predator, *Cichlasoma managuense* (Gunther), *Aquaculture*, 14, 383, 1978.
415. McGinty, A. S., Population dynamics of peacock bass, *Cichla ocellaris* and *Oreochromis* (*Tilapia*) *niloticus* in fertilized ponds, in *Int. Symp. Tilapia in Aquaculture*, Fishelson, L. and Yaron, Z., Eds., Tel Aviv University, Tel Aviv, 1984, 86.
416. Pruginin, Y., Report to the Government of Uganda on the Experimental Fish Culture Project in Uganda, 1962—1964, FAO/UNDP (Technical Assistance) Reports on Fisheries, TA Reports, Food and Agriculture Organization, Rome, 1960, 1965, 1.
417. Pruginin, Y., Report to the Government of Uganda on the Experimental Fish Culture Project in Uganda, 1965—1966, FAO/UNDP (Technical Assistance) Reports on Fisheries, TA Reports, Food and Agriculture Organization, Rome, 2446, 1967, 1.
418. Meschkat, A., The status of warm-water fish culture in Africa, *FAO Fish. Rep.*, 44, 88, 1967.
419. Chervinski, J., Sea bass, *Dicentrarchus labrax* (Pisces, Serranidae) a "policefish" in freshwater ponds and its adaptability to saline conditions, *Bamidgeh*, 26, 110, 1974.
420. Swingle, H. S., Comparative evaluation of two tilapias as pondfishes in Alabama, *Trans. Am. Fish. Soc.*, 89, 142, 1960.
421. Sarig, S. and Arieli, Y., Growth capacity of tilapia in intensive culture, *Bamidgeh*, 32, 57, 1980.
422. Chervinski, J. and Rothbard, S., An aid in manually sexing *Tilapia*, *Aquaculture*, 26, 389, 1982.
423. Chen, F. Y., Preliminary studies on the sex-determining mechanism of *Tilapia mossambica* Peters and *T. hornnorum* Trewavas, *Verh. Int. Ver. Limnol.*, 17, 719, 1969.
424. Hickling, C. F., The Malacca *Tilapia* hybrids, *J. Genet.*, 57, 1, 1960.
425. Guerrero, R. D. and Caguan, A. G., Culture of male *Tilapia mossambica* produced through artificial sex reversal, in *Advances in Aquaculture*, Pillay, T. V. R. and Dill, W. A., Eds., Fishing News Books Ltd., Farnham, Surrey, England, 1979, 166.
426. Hsiao, S. M., Hybridization of *Tilapia mossambica*, *T. nilotica*, *T. aurea* and *T. zillii* — a preliminary report, *China Fish. Mon.* (*Taipei*), 332, 3, 1980.

427. Lee, J. C., Reproduction and Hybridization of Three Cichlid Fishes, *Tilapia aurea* (Steindachner), *T. hornorum* (Trewavas) and *T. nilotica* (Linnaeus) in Aquaria and Plastic Pools, Ph.D. dissertation, Auburn University, Auburn, Ala., 1979, 1.

428. Pinto, L. G., Hybridization between species of tilapia, *Trans. Am. Fish. Soc.,* 111, 481, 1982.

429. Pruginin, Y., Rothbard, S., Wohlfarth, G., Halevy, A., Moav, R., and Hulata, G., All-male broods of *Tilapia nilotica* x *T. aurea* hybrids, *Aquaculture,* 6, 11, 1975.

430. Jalabert, B., Kammacher, P., and Lessent, P., Sex determination in *Tilapia machrochir* x *Tilapia nilotica* hybrids. Investigations on sex ratios in first generation x parent crossings, *Biochim. Biophys.,* 11, 155, 1971.

431. Fishelson, L., Hybrids of two species of fishes in the genus *Tilapia* (Cichlidae, Teleostei), *Fishermen's Bull.,* 4(2), 14, 1962.

432. Peters, H. M., Untersuchungen zum Problem des angeborenen Verhaltens, *Naturwissenschaften,* 50, 677, 1963.

433. Bauer, J., Vergleichende Untersuchungen zum Kontaktverhalten verschiedener Arten der Gattung *Tilapia* (Cichlidae, Pisces) und ihrer Bastarde, *Z. Tierpsychol.,* 25, 22, 1968.

434. Wohlfarth, G. W. and Hulata, G., *Applied Genetics of Tilapias,* ICLARM Studies and Reviews, Vol. 6, International Center for Living Aquatic Resources Management, Manila, Philippines, 1983, 1.

435. Hopkins, K. D., Shelton, W. L., and Engle, C. R., Estrogen sex-reversal of *Tilapia aurea, Aquaculture,* 18, 263, 1979.

436. Jensen, G. L. and Shelton, W. L., Effects of estrogens on *Tilapia aurea:* implications for production of monosex genetic male tilapia, *Aquaculture,* 16, 233, 1979.

437. Clemens, H. P. and Inslee, T., The production of unisexual broods of *Tilapia mossambica* sex-reversed with methyltestosterone, *Trans. Am. Fish. Soc.,* 97, 18, 1968.

438. Guerrero, R. D., Culture of male *Tilapia mossambica* produced through artificial sex-reversal, in *Advances in Aquaculture,* Pillay, T. V. R., and Dill, W. A., Eds., Fishing News Books Ltd., Farnham, Surrey, England, 1979, 166.

439. Jalabert, B., Moreau, J., Planquette, P., and Billard, R., Determinisme de sexe chez *Tilapia macrochir* et *Tilapia nilotica.* Action de la methyltestosterone dans l'alimentation des alevins sur la differentiation sexuelle; proportion des sexes dans la descendance des males "inverses", *Ann. Biol. Anim. Biochim. Biophys.,* 14, 729, 1974.

440. Guerrero, R. D. and Abella, T. A., Induced sex-reversal of *Tilapia nilotica* with methyltestosterone, *Fish. Res. J. Philipp.,* 1, 46, 1976.

441. Tayamen, M. M. and Shelton, W. L., Inducement of sex reversal in *Sarotherodon niloticus* (Linnaeus), *Aquaculture,* 14, 349, 1978.

442. Guerrero, R. D., III, Use of androgens for the production of all-male *Tilapia aurea* (Steindachner), *Trans. Am. Fish. Soc.,* 104, 342, 1975.

443. Shelton, W. L., Rodriguez-Guerrero, D., and Lopez-Marcias, J., Factors affecting androgen sex reversal of *Tilapia aurea, Aquaculture,* 25, 59, 1981.

444. Marek, M., Revision of supplementary feeding tables for pond fish, *Bamidgeh,* 27, 57, 1975.

445. Spataru, P. and Zorn, M., Food and feeding habits of *Tilapia aurea* (Steindachner) (Cichlidae) in Lake Kinneret (Israel), *Aquaculture,* 13, 67, 1978.

446. Spataru, P. and Gophen, M., A comparative study of the food and feeding habits of *Sarotherodon galilaeus* and *Oreochromis aureus* (Cichlidae) from Lake Kinneret (Abstr.), in Int. Symp. *Tilapia* in Aquaculture, Nazareth, Israel, May 8 to 13, 1983, 28.

447. Moriarty, C. M. and Moriarty, D. J. W., Quantitative estimation of the daily ingestion of phytoplankton by *Tilapia nilotica* and *Haplochromis nigripinnis* in Lake George, Uganda, *Proc. R. Soc. Lond. Ser. B,* 184, 299, 1973.

448. Bowen, S. H., A nutritional constraint in detritivory by fishes: the stunted population of *Sarotherodon mossambicus* in Lake Sigaya, South Africa, *Ecol. Monogr.,* 49, 17, 1979.

449. Bowen, S. H., Detrital nonprotein amino acids are the key to rapid growth of tilapia in Lake Valencia, Venezuela, *Science,* 207, 1216, 1980.

450. Man, H. S. H. and Hodgkiss, I. J., Studies on the ichthyo-fauna in Plover Cove Reservoir, Hong Kong: feeding and food relations, *J. Fish Biol.,* 11, 1, 1977.

451. Munro, J. L., The food of a community of East African freshwater fishes, *J. Zool.,* 151, 389, 1967.

452. Naik, I. U., Studies on *Tilapia mossambica* Peters in Pakistan, *Agric. Pak.,* 24, 47, 1973.

453. Weatherly, A. H. and Cogger, B. M. G., Fish culture: problems and prospects, *Science,* 197, 427, 1977.

454. Spataru, P., The feeding habits of *Tilapia galilaea* (Artedi) in Lake Kinneret (Israel), *Aquaculture,* 9, 47, 1976.

455. Abdel-Malek, S. A., Food and feeding habits of some Egyptian fishes in Lake Quarun. I. *Tilapia zillii* (Gerv.). B. According to different length groups, *Bull. Inst. Oceanogr. Fish. Cairo,* 2, 204, 1972.

456. Buddington, R. K., Digestion of an aquatic macrophyte by *Tilapia zillii, J. Fish. Biol.,* 15, 449, 1979.

457. Davis, A. T. and Stickney, R. R., Growth responses of *Tilapia aurea* to dietary protein quality and quantity, *Trans. Am. Fish. Soc.*, 107, 479, 1978.

458. Jauncey, K., The effects of varying dietary protein level on the growth, food conversion, protein utilization and body composition of juvenile tilapias (*Sarotherodon mossambicus)*, *Aquaculture*, 27, 43, 1982.

459. Mazid, M. A., Tanaka, Y., Katayama, T., Rahman, M. A., Simpson, K. L., and Chichester, C. O., Growth response of *Tilapia zillii* fingerlings fed iso-caloric diets with variable protein levels, *Aquaculture*, 18, 115, 1979.

460. Newman, M. W., Huezo, H. E., and Hughes, D. G., The response of all-male tilapia hybrids to four levels of protein in isocaloric diets, *Proc. World Maricult. Soc.*, 10, 788, 1979.

461. Winfree, R. A. and Stickney, R. R., Effects of dietary protein and energy on growth, feed conversion efficiency and body composition of *Tilapia aurea*, *J. Nutr.*, 111, 1001, 1981.

462. Jackson, A. J. and Capper, B. S., Investigations into the requirements of the tilapia *Sarotherodon mossambicus* for dietary methionine, lysine and arginine in semi-synthetic diets, *Aquaculture*, 29, 289, 1982.

463. Jauncey, K., Tacon, A. G. J., and Jackson, A. J., The quantitative essential amino acid requirements of *Oreochromis* (= *Sarotherodon) mossambicus*, in Int. Symp. *Tilapia in Aquaculture*, Fishelson, L. and Yaron, Z., Eds., Tel Aviv University, Tel Aviv, 1984, 328.

464. Viola, S. and Arieli, Y., Nutrition studies with tilapia (*Sarotherodon*). I. Replacement of fishmeal by soybeanmeal in feeds for intensive tilapia culture, *Bamidgeh*, 35, 9, 1983.

465. Jackson, A. J., Capper, B. S., and Matty, J. A., Evaluation of some plant proteins in complete diets for the tilapia *Sarotherodon mossambicus, Aquaculture*, 27, 97, 1982.

466. Tacon, A. G. J., Jauncey, K., Falaye, A. E., and Pantha, M. B., The use of meat and bone meal, hydrolysed feathermeal and soybean meal in practical fry and fingerling feeds for *Oreochromis niloticus*, in Int. Symp. *Tilapia in Aquaculture*, Fishelson, F. and Yaron, Z., Eds., Tel Aviv University, Tel Aviv, 1984, 356.

467. Appler, H. N. and Jauncey, K., The utilization of a filamentous green alga (*Cladophora glomerata* (L) Kutzin) as a protein source in pelleted feeds for *Sarotherodon (Tilapia) niloticus* fingerlings, *Aquaculture*, 30, 21, 1983.

468. Moriarty, D. J. W., The physiology of digestion of blue-green algae in the cichlid fish, *Tilapia nilotica, J. Zool. Lond.*, 171, 25, 1973.

469. Bayne, D. R., Dunseth, D., and Ramirios, C. G., Supplemental feeds containing coffee pulp for rearing *Tilapia* in Central America, *Aquaculture*, 7, 133, 1976.

470. Cruz, E. M. and Laudencia, I. L., Screening of feedstuffs as ingredients in the rations of Nile tilapia, *Kalikasan, Philipp. J. Biol.*, 7, 159, 1978.

471. Kohler, C. C. and Pagan-Font, F. A., Evaluations of rum distillation wastes, pharmaceutical wastes and chicken feed for rearing *Tilapia aurea* in Puerto Rico, *Aquaculture*, 14, 339, 1978.

472. Guerrero, R. D., III, Studies on the feeding of *Tilapia nilotica* in floating cages, *Aquaculture*, 20, 169, 1980.

473. Gophen, M., Food sources, feed behaviour and growth rates of *Sarotherodon galilaeus* (Linnaeus) fingerlings, *Aquaculture*, 20, 101, 1980.

474. Stickney, R. R., Simmons, H. B., and Rowland, L. O., Growth responses of *Tilapia aurea* to feed supplemented with dried poultry waste, *Tex. J. Sci.*, 28, 93, 1977.

475. Degani, G., Dosoretz, C., Levanon, D., Marchaim, U., and Perach, Z., Feeding *Sarotherodon aureus* with fermented cow manure, *Bamidgeh*, 34, 119, 1982.

476. Anderson, J., Jackson, A. J., and Matty, J. A., Effects of purified carbohydrates and fibre on the growth of the *Orechromis (Tilapia) niloticus* (Abstr.), in Int. Symp. Tilapia in Aquaculture, Nazareth, Israel, May 8 to 13, 1983, 80.

477. Stickney, R. R., Cellulase activity in the stomachs of freshwater fishes from Texas, *Proc. S.E. Assoc. Game and Fish Comm.*, 26, 282, 1976.

478. Kanazawa, A., Teshima, S., Sakamoto, M., and Awal, M. A., Requirements of *Tilapia zillii* for essential fatty acids, *Bull. Jpn. Soc. Fish.*, 46, 1353, 1980.

479. Stickney, R. R. and McGeachin, R. B., Responses of *Tilapia aurea* to semipurified diets of differing fatty acid composition, in *Proc. Conf. Tilapia in Aquaculture*, Fishelson, F. and Yaron, Z., Eds., Tel Aviv University, Tel Aviv, 1984, 346.

480. Stickney, R. R. and McGeachin, R. B., Effects of dietary lipid quality on growth and food conversion of *Tilapia aurea, Proc. S.E. Assoc. Fish and Wildl. Agen.*, in press.

481. Stickney, R. R., McGeachin, R. B., Lewis, D. H., Marks, J., Riggs, A., Sis, R. F., Robinson, E. H., and Wurts, W., Response of *Tilapia aurea* to dietary vitamin C, *Proc. World Maricult. Soc.*, 15, 179, 1984.

482. Oyetayo, A. S., Thornburn, C. C., and Matty, A. J., Effects of dietary ascorbic acid on growth parameters of *Tilapia Oreochromis mossambicus* fry (Abstr.), in Int. Symp. Tilapia in Aquaculture, Nazareth, Israel, May 8 to 13, 1983, 75.

483. Lovell, R. T. and Limsuwan, T., Intestinal synthesis and dietary nonessentiality of vitamin B$_{12}$ in *Tilapia nilotica, Trans. Am. Fish. Soc.,* 111, 485, 1982.

484. Matsuno, T., Katsuyama, M., Iwahashi, M., Koike, T., and Okada, M., Intensification of color of red *Tilapia* with lutein, rhodoxanthin and spirulina, *Bull. Jpn. Soc. Sci. Fish.,* 46, 479, 1980.

485. Roberts, R. J. and Sommerville, C., Diseases of tilapias, in *The Biology and Culture of Tilapias,* Pullin, R. S. V., and Lowe-McConnell, R. H., Eds., ICLARM Conf. Proc., Vol. 7, International Center for Living Aquatic Resource Management, Manila, Philippines, 1982, 247.

486. Swingle, H. S. and Smith, E. V., Management of Farm Fish Ponds, Agric. Exp. Sta. Ala. Polytech. Inst., Auburn, 1938, 1.

487. Swingle, H. S. and Smith, E. V., The Management of Farm Fish Ponds, Bull. 254, Ala. Agric. Exp. Sta., Auburn, 1942, 1.

488. Swingle, H. S., Improvement of fishing in old ponds, *Trans. North Am. Wildl. Conf.,* 10, 299, 1945.

489. Swingle, H. S., Relationships and Dynamics of Balanced and Unbalanced Fish Populations, Bull. 274, Ala. Polytech. Inst. Agric. Exp. Sta., Auburn, 1950, 1.

490. Davies, W. D., Shelton, W. L., and Malvestuto, S. P., Prey-dependent recruitment of largemouth bass: a conceptual model, *Fisheries,* 7(6), 12, 1982.

491. Lawrence, J. M., Estimated size of various forage fishes largemouth bass can swallow, *Proc. S.E. Assoc. Game Fish Comm.,* 11, 220, 1958.

492. Williamson, J. H., Comparing training success of two strains of largemouth bass, *Prog. Fish-Cult.,* 45, 3, 1983.

493. Moorman, R. B., Reproduction and growth of fishes in Marion Co., Iowa farm ponds, *Iowa St. Coll. J. Sci.,* 32, 71, 1957.

494. Heidinger, R. C., Synopsis of Biological Data on the Largemouth Bass *Micropterus salmoides* (Lacepede) 1802, FAO Fisheries Synopsis No. 115, Food and Agriculture Organization, Rome, 1976, 1.

495. Williamson, J. H., unpublished data.

496. Snow, J. R., Jones, R. O., and Rogers, W. A., Training Manual of Warmwater Fish Culture, Bureau of Sport Fisheries and Wildlife, Washington, D.C., 1964, 1.

497. Parker, W. D., Preliminary studies on sexing adult largemouth by means of an external characteristic, *Prog. Fish-Cult.,* 36, 55, 1971.

498. Driscoll, D. P., Sexing the largemouth bass with an otoscope, *Prog. Fish-Cult.,* 31, 183, 1969.

499. Snow, J. R., Fecundity of largemouth bass, *Micropterus salmoides* (Lacepede), receiving artificial food, *Proc. S.E. Assoc. Game Fish Comm.,* 24, 550, 1971.

500. Kramer, R. H. and Smith, L. L., Jr., Formation of year classes in largemouth bass, *Trans. Am. Fish. Soc.,* 91, 29, 1962.

501. Swingle, H. S., Appraisal of methods of fish population study. IV. Determination of balance in farm fish ponds, *Trans. North Am. Wildl. Conf.,* 21, 298, 1956.

502. Kramer, R. H. and Smith, L. L., Jr., First-year growth of the largemouth bass, *Micropterus salmoides,* and some related ecological factors, *Trans. Am. Fish. Soc.,* 89, 222, 1960.

503. Snow, J. R., hatchery propagation of the black bass in *Black Bass Biology and Management,* Clapper, J. E., Ed., Sport Fishing Institute, Washington, D.C., 344, 1975.

504. Anonymous, Manual of Fish Culture, U.S. Fish Commission, Washington, D.C., 1900, 1.

505. Snow, J. R., Results of further experiments on rearing largemouth bass fingerlings under controlled conditions, *Proc. S.E. Assoc. Game Fish Comm.,* 17, 303, 1963.

506. Snow, J. R., Some progress in the controlled culture of the largemouth bass, *Micropterus salmoides* (Lacepede), *Proc. S.E. Assoc. Game Fish Comm.,* 22, 380, 1968.

507. Bishop, H., Largemouth bass culture in the southwest, in Proc. N. Central Warmwater Fish Culture Workshop, Ames, Iowa, 1968, 24.

508. Jackson, U. T., Controlled spawning of largemouth bass, *Prog. Fish-Cult.,* 41, 90, 1979.

509. Carlson, A. R. and Hale, J. G., Successful spawning of largemouth bass *Micropterus salmoides* (Lacepede) under laboratory conditions, *Trans. Am. Fish. Soc.,* 101, 539, 1972.

510. Carlson, A. R., Induced spawning of largemouth bass, *Micropterus salmoides* (Lacepede), *Trans. Am. Fish. Soc.,* 102, 442, 1973.

511. McCraren, J. P., Feeding young bass, *Farm Pond Harvest,* 9(3), 10, 1975.

512. Huet, M., *Textbook on Fish Culture; Breeding and Cultivation of Fish,* Fishing News (Books) Ltd., London 1970, 1.

513. Chastain, G. A. and Snow, J. R., Nylon mats as spawning sites for largemouth bass, *Micropterus salmoides* (Lacepede), *Proc. S.E. Assoc. Game Fish Comm.,* 19, 405, 1966.

514. White, B. L., Culture of Florida largemouth bass *Micropterus salmoides floridanus,* in Midwest Black Bass Culture, Hutson, P. L. and Lillie, J., Eds., Texas Parks and Wildlife Department, Austin, and Kansas Fish and Game Commission, Pratt, 1982, 146.

515. Snow, J. R., Culture of largemouth bass, in Report of the 1970 Workshop on Fish Feed Technology and Nutrition, Resource Publ. U.S. Bur. Sport Fish. Wildl. 102, Washington, D.C., 1970, 86.

516. Wilbur, R. L. and Langford, F., Use of human chorionic gonadotropin (HCG) to promote gametic production in male and female largemouth bass, *Proc. S.E. Assoc. Game Fish Comm.*, 28, 242, 1975.

517. Anon., Spawn black bass in concrete raceway, *Aquaculture Mag.*, 10(6A), 46, 1984.

518. Timmons, T. J., Shelton, W. L., and Davies, W. D., Gonad development, fecundity, and spawning season of largemouth bass in newly impounded West Point Reservoir, Alabama-Georgia, Tech. Pap. 100, U.S. Fish and Wildl. Serv., Washington, D.C., 1980, 1.

519. Chew, R. L., The failure of largemouth bass, *Micropterus salmoides floridanus* (LeSueur), to spawn in eutrophic, over-crowded environments, *Proc. S.E. Assoc. Game Fish Comm.*, 26, 306, 1974.

520. Stevens, R. E., Hormonal Relationships Affecting Maturation and Ovulation in Largemouth Bass (*Micropterus salmoides* Lacepede), Ph.D. dissertation, North Carolina State University, Raleigh, 1970, 1.

521. Snow, J. R., Controlled culture of largemouth bass fry, *Proc. S.E. Assoc. Game Fish Comm.*, 26, 392, 1972.

522. Badenhuizen, T. R., Effect of Incubation Temperature on Mortality of Embryos of the Largemouth Bass, *Micropterus salmoides* (*Lacepede*), M.S. thesis, Cornell University, Ithaca, New York, 1969, 1.

523. Merriner, J. V., Development of intergenetic centrarchid hybrid embryos, *Trans. Am. Fish. Soc.*, 100, 611, 1971.

524. Johnson, P. M., The embryonic development of the swim bladder of the largemouth black bass, *Micropterus salmoides salmoides* (Lacepede), *J. Morphol.*, 93, 45, 1953.

525. Meehan, O. L., A method for the production of largemouth bass on natural food in fertilized ponds, *Prog. Fish-Cult.*, 47, 1, 1939.

526. Anon., *Prog. Fish-Cult.*, 1951.

527. Anderson, J. R., Glass V-trap, *Prog. Fish-Cult.*, 36, 53, 1974.

528. Swanson, R., 1982 bass propagation at the Las Animas Fish Hatchery, in *Midwest Black Bass Culture*, Hutson, P. L. and Lillie, J., Eds., *Texas Parks and Wildlife Department, Austin, and Kansas Fish and Game Commission, Pratt*, 1982, 25.

529. Bowling, C. W., Rudledge, W. P., and Geiger, J. G., Evaluation of ryegrass cover crops in rearing ponds for Florida largemouth bass, *Prog. Fish-Cult.*, 46, 55, 1984.

530. Geiger, J. G., A review of pond zooplankton production and fertilization for the culture of larval and fingerling striped bass, *Aquaculture*, 35, 353, 1983.

531. Carmichael, G. J. and Tomasso, J. R., Use of formalin to separate tadpoles from largemouth bass fingerlings after harvesting, *Prog. Fish-Cult.*, 45, 105, 1983.

532. Snow, J. R., An exploratory attempt to rear largemouth black bass fingerlings in a controlled environment, *Proc. S.E. Assoc. Game Fish Comm.*, 14, 253, 1960.

533. Langlois, T. H., The problem of the efficient management of hatcheries used in the production of pond fishes, *Trans. Am. Fish. Soc.*, 61, 106, 1931.

534. Snow, J. R., The Oregon Moist Pellet as a diet for largemouth bass, *Prog. Fish-Cult.*, 30, 235, 1968.

535. Nagle, T., Rearing Largemouth Bass Yearlings on Artificial Diets, Division Wildlife In Service Note 335, Ohio Department Natural Resources, 1976, 1.

536. Willis, D. W. and Flickinger, S. A., Survey of private and government hatchery success in raising largemouth bass, *Prog. Fish-Cult.*, 42, 232, 1980.

537. Westers, H. and Pratt, K. M., The rational design of fish hatchery for intensive culture based on metabolic characteristics, *Prog. Fish-Cult.*, 39, 157, 1977.

538. Nelson, J. T., Bowker, R. G., and Robinson, J. D., Rearing pellet-fed largemouth bass in a raceway, *Prog. Fish-Cult.*, 36, 108, 1974.

539. McCraren, J. P., Hatchery production of advanced largemouth bass fingerlings, *Proc. S.E. Assoc. Game Fish Comm.*, 54, 260, 1974.

540. Snow, J. R., Brewer, D., and Wright, C. R., Plastic bags for shipping sac fry of largemouth bass, *Prog. Fish-Cult.*, 40, 13, 1978.

541. Wedemeyer, G., The role of stress in disease resistance of fishes, in Snieszko, S. F., Ed., *A Symp. on Diseases of Fishes and Shellfishes*, Spec. Publ. No. 5, American Fisheries Society, Bethesda, Md., 30, 1970.

542. Lewis, S. D., The Effect of Salt Solutions on Osmotic Changes Associated with Surface Damage to the Golden Shiner, *Notemigonus chrysoleucas*, Diss. Abstr., Part 13, 1971, 6346.

543. Mazeud, M. M., Mazeud, F., and Donaldson, E. M., Primary and secondary effect of stress in fish: some new data with a general review, *Trans. Am. Fish. Soc.*, 106, 201, 1977.

544. Selye, H., Stress and the general adaptation syndrome, *Br. Med. J.*, 1, 1383, 1950.

545. Schreck, C. B., Stress and Rearing of Salmonids, Oregon State Univ. Agric. Exp. Sta. Tech. Pap. No. 5911, 1982, 1.

546. Pic, P., Mayer-Gostan, N., and Maetz, J., Branchial effects of epinephrine in the seawater-adapted mullet. I. Water permeability, *Am. J. Physiol.*, 226, 698, 1974.

547. Norton, V. M. and Davis, K. B., Effect of abrupt change in the salinity of the environment on the plasma electrolytes, urine volume, and electrolyte excretion in channel catfish, *Ictalurus punctatus, Comp. Biochem. Physiol.*, 56A, 425, 1977.

548. Grant, N., Metabolic effects of adrenal glucocorticoid hormones, in *The Adrenal Cortex*, Eisenstein, A. B., Ed., Little, Brown, Boston, 269, 1967.

549. Carmichael, G. J., Tomasso, J. R., Simco, B. A., and Davis, K. B., Characterization and alleviation of stress associated with hauling largemouth bass, *Trans. Am. Fish. Soc.*, 113, 778, 1984.

550. Carmichael, G. J., Tomasso, J. R., Simco, B. A., and Davis, K. B., Confinement and water quality-induced stress in largemouth bass, *Trans. Am. Fish. Soc.*, 113, 767, 1984.

551. Strange, R. J. and Schreck, C. B., Anesthetic and handling stress on survival and the cortisol response in yearling chinook salmon (*Oncorhynchus tshawytscha*), *J. Fish. Res. Bd. Can.*, 35, 345, 1978.

552. Tomasso, J. R., Davis, K. B., and Parker, N. C., Plasma corticosteroid and electrolyte dynamics of hybrid striped bass (white bass x striped bass) during netting and hauling, *Proc. World Maricult. Soc.*, 11, 303, 1980.

553. Collins, J. L. and Hulsey, A. M., Hauling mortality of threadfin shad reduced with MS-222 and salt, *Prog. Fish-Cult.*, 25, 105, 1963.

554. Hattingh, J., LeRoux, Fourie, F., and Van Vuren, J. H. J., The transport of freshwater fish, *J. Fish Biol.*, 7, 447, 1975.

555. Coetzee, N. and Hattingh, J., Effects of sodium chloride on the freshwater fish, *Labeo capensis*, during and after transportation, *Zool. Afr.*, 12, 244, 1977.

556. Murai, T., Andrews, J. W., and Muller, J. W., Fingerling American shad: effects of valium, MS-222, and sodium chloride on handling mortality, *Prog. Fish-Cult.*, 41, 27, 1979.

557. Leitritz, E. and Lewis, R. C., Trout and salmon culture (hatchery methods), *Fish Bull. Calif. Dept. Fish Game*, 164, 1976, 1.

558. Smith, C. E., United States Fish and Wildlife Service, Manual of Fish Culture, Section G, Fish Transportation, U.S. Government Printing Office, Washington, D.C., 1978, 1.

559. Inslee, T. D., Increased production of smallmouth bass fry, in *Black Bass Biology and Management*, Clepper, H., Ed., Sport Fishing Institute, Washington, D.C., 1975, 357.

560. Carmichael, G. J., Wedemeyer, G. A., McCraren, J. P., and Millard, J. L., Physiological effects of handling and hauling stress on smallmouth bass, *Prog. Fish-Cult.*, 45, 110, 1983.

561. Flickinger, S. A. and Williamson, J. H., Information presented at the U.S. Fish and Wildlife Service Warmwater Fish Culture Workshop, San Marcos National Fish Hatchery and Technology Center, San Marcos, Texas., 1984.

562. Higginbotham, B. J., Noble, R. L., and Rudd, A., Culture techniques of forage species commonly utilized in Texas waters, mimeo, undated, 1.

563. Blosz, J., Fish production program, 1947, in the southeast, *Prog. Fish-Cult.*, 10, 61, 1948.

564. Surber, E. W., Chemical control agents and their effects on fish, *Prog. Fish-Cult.*, 10, 125, 1948.

565. Davis, H. S., *Culture and Diseases of Game Fishes*, University of California Press, Los Angeles, 1953, 1.

566. McComish, T. S., Sexual differentiation of bluegills by the urogenital opening, *Prog. Fish-Cult.*, 30, 28, 1968.

567. Lewis, W. M. and Heidinger, R. C., Use of hybrid sunfishes in the management of small impoundments, in *New Approaches to the Management of Small Impoundments*, Novinger, G. D. and Dillard, J. G., Ed., North Central Div. Am. Fish. Soc., Spec. Publ. No. 5, 1978, 104.

568. Lewis, W. M. and Heidinger, R. C., Supplemental feeding of hybrid sunfish populations, *Trans. Am. Fish. Soc.*, 100, 619, 1971.

569. Ming, A., A Review of the Literature on Crappies in Small Impoundments, Mo. Dept. Cons. Div. Fish. Res. Sec. Proj. No. F-1-R-20, Study No. I-15, No. 1, 1971, 1.

570. Smeltzer, J. F., Culture, Handling, and Feeding Techniques for Black Crappie Fingerlings, M.S. thesis, Colorado State University, Fort Collins, 1981, 1.

571. Leary, J. L., Propagation of crappie and catfish, *Trans. Am. Fish. Soc.*, 38, 143, 1909.

572. Davis, H. S. and Wiebe, A. H., Experiments in the Culture of Black Bass and Other Pondfish, Doc. No. 1085, U.S. Bureau Fisheries, Washington, D.C., 1930, 1.

573. Culler, C. F., Notes on warm-water fish culture, *Prog. Fish-Cult.*, 36, 19, 1938.

574. Harper, D. C., Crappie and calico bass culture in Texas, *Prog. Fish-Cult.*, 38, 12, 1938.

575. Mraz, D. and Cooper, E. L., Reproduction of carp, largemouth bass, bluegill, and black crappies in small rearing ponds, *J. Wildl. Man.*, 21, 127, 1957.

576. Siefert, R. E., Reproductive behavior, incubation, and mortality of eggs, and postlarval food selection in the white crappie, *Trans. Am. Fish. Soc.*, 97, 252, 1968.

577. Siefert, R. E., Carlson, A. R., and Herman, L. J., Effects of reduced oxygen concentrations on the early life stages of mountain whitefish, smallmouth bass and white bass, *Prog. Fish-Cult.*, 36, 186, 1974.

578. Dudley, R. G. and Eipper, A. W., Survival of largemouth bass embryos at low dissolved oxygen concentrations, *Trans. Am. Fish. Soc.*, 104, 122, 1975.

579. Stewart, N. C., Shumway, D. L., and Doudoroff, P., Influence of oxygen concentration on the growth of juvenile largemouth bass, *J. Fish. Res. Bd. Can.*, 24, 475, 1967.

580. Warren, C. E., Doudoroff, P., and Shumway, D. L., Development of Dissolved Oxygen Criteria for Freshwater Fish, USEPA-R3-73-019, U.S. Environmental Protection Agency, Washington, D.C., 1973, 1.

581. Moss, D. D. and Scott, D. C., Dissolved oxygen requirements of three species of fish, *Trans. Am. Fish. Soc.*, 90, 377, 1961.

582. Whitmore, C. M., Warren, C. E., and Doudoroff, P., Avoidance reactions of salmonid and centrarchid fishes to low oxygen concentrations, *Trans. Am. Fish. Soc.*, 89, 17, 1960.

583. Kelly, J. W., Effects of incubation temperature on survival of largemouth bass eggs, *Prog. Fish-Cult.*, 30, 159, 1968.

584. Coutant, C. C. and DeAngelis, D. L., Comparative temperature-dependent growth rates of largemouth and smallmouth bass fry, *Trans. Am. Fish. Soc.*, 112, 416, 1983.

585. Lemke, A., Optimum temperature for growth of juvenile bluegills, *Prog. Fish-Cult.*, 39, 55, 1977.

586. Wrenn, W. B., Effects of elevated temperature on growth and survival of smallmouth bass, *Trans. Am. Fish. Soc.*, 109, 617, 1980.

587. Tebo, L. B. and McCoy, E. G., Effect of sea-water concentration on the reproduction and survival of largemouth bass, *Prog. Fish-Cult.*, 26, 99, 1964.

588. Ellgard, E. G. and Gilmore, J. Y., III, Effects of different acids on the bluegill sunfish, *Lepomis macrochirus* Rafinesque, *J. Fish Biol.*, 25, 133, 1984.

589. Ultsch, G., Oxygen consumption as a function of pH in three species of freshwater fishes, *Copeia*, 1978, 272, 1978.

590. Emery, R. M. and Welch, E. B., The toxicity of Alkaline Solutions of Ammonia to Juvenile Bluegill Sunfish (*Lepomis macrochirus* Raf.), Water Quality Branch, Division of Health and Safety, Tennessee Valley Authority, Chatanooga, 1969, 1.

591. Roseboom, D. P. and Richey, D. L., Acute Toxicity of Residual Chlorine and Ammonia to Native Illinois Fishes, Report of Investigation 85, Illinois State Water Survey, Urbana, 1977, 1.

592. Carmichael, G. J. and Tomasso, J. R., unpublished data.

593. Wedemeyer, G. A. and Yasutake, W. T., Prevention and treatment of nitrite toxicity in juvenile steelhead trout (*Salmo gairdneri*), *J. Fish. Res. Bd. Can.*, 35, 822, 1978.

594. Huey, D. W., Wooten, M. C., Freeman, L. A., and Beitinger, T. L., Effect of pH and chloride on nitrite-induced lethality in bluegill (*Lepomis macrochirus*), *Bull. Environ. Contam. Toxicol.*, 28, 3, 1982.

595. Palachek, R. M. and Tomasso, J. R., Toxicity of nitrite to channel catfish, tilapia, and largemouth bass: evidence for a nitrite exclusion mechanism in largemouth bass, *Can. J. Fish. Aquat. Sci.*, in press, 1984.

596. Huisman, E. A., Richter, C. J. J., and Hogendorn, H., *Visteelt*, Agricultural University of Wageningen, The Netherlands, 1980, 1.

597. Clark, J. H., *Variability of Northern Pike Pond Culture Production*, M.S. thesis, Colorado State University, Fort Collins, 1974, 1.

598. Sorenson, L., Buss, K., and Bradford, D., The artificial propagation of esocid fishes in Pennsylvania, *Prog. Fish-Cult.*, 28, 133, 1966.

599. Ragan, J., Steinwand, T., and VanEeckhout, G., A synopsis of a muskellunge questionnaire survey formulated for the international symposium on muskellunge, in *Int. Symp. Muskellunge*, LaCrosse, Wisconsin, 1984, in press.

600. Colesante, R. J., Improvement of Esocid Culture Techniques, Report of 1976 studies, New York Department of Environmental Conservation, 1977, 1.

601. Bender, T. R., Jr., Mudrak, V. A., and Hood, S. E., Final Report on Intensive Culture of Tiger Muskellunge Using Prepared Diets, Pennsylvania Fish Commission, 1979, 1.

602. Pecor, C. H. and Humphrey, R., Tiger Muskies at Michigan's Platte River Hatchery in 1980, 82, and 83, Michigan Department of Natural Resources, 1984, 1.

603. Graff, D. R. and Sorenson, L., The successful feeding of a dry diet to esocids, *Prog. Fish-Cult.*, 32, 31, 1970.

604. Timmermans, G. A., Culture of Fry and Fingerlings of Pike, *Esox lucius*, EIFAC/T35, Suppl. 1, 177, 1979.

605. Phillips, R. A. and Graveen, W. J., A domestic muskellunge broodstock, *Prog. Fish-Cult.*, 35, 176, 1973.

606. Klingbiel, J. H., Culture of purebred muskellunge, *Int. Symp. Muskellunge*, LaCrosse, Wis., 1984, in press.

607. Jennings, T., Hatchery Branch Production Report, Iowa Conservation Commission, 1982, 1.

179

608. de Montalembert, G., Bry, C., and Billard, R., Control of reproduction in northern pike, in *Selected Coolwater Fishes of North America*, Special Publ. No. 11, Kendall, R. L., Ed., American Fisheries Society, Washington, D.C., 217, 1978.

609. Hassler, T. J., Effect of temperature on survival of northern pike embryos and yolk-sac larvae, *Prog. Fish-Cult.*, 44, 174, 1982.

610. Huisman, E. A., Skjervold, H., and Richter, C. J. J., Aspects of Fish Culture and Fish Breeding, Miscellaneous Papers 13, Agricultural University of Wageningen, The Netherlands, 1976, 1.

611. Johnson, L. J., Pond Culture of Muskellunge in Wisconsin, Wisconsin Conservation Department Tech. Bull., 1958, 17.

612. Klingbiel, J. H., Ed., Warmwater Stocking and Propagation Audit, Wisconsin Department of Natural Resources, 1983, 1.

613. Menz, F. C., An Economic Study of the New York State Muskellunge Fishery, New York Sea Grant Institute, Albany, 1, 37, 1978.

614. Hiner, L. E., Propagation of northern pike, *Trans. Am. Fish. Soc.*, 90, 298, 1961.

615. Braum, E., Experimentelle Untersuchungen zur ersten Nahrungsaufnahme und Biologie am Jungfischen von Blaufelchen (*Coregonus wartmanni*, Bloch), Weissfelchen (*Coregonus fera*, Jurine) und Hechten (*Esox lucius*, L.), *Arch. Hydrobiol.*, Suppl. 28, 183, 1964.

616. Jorgensen, W., Iowa culture of muskellunge on artificial diets, *Int. Symp. Muskellunge*, LaCrosse, Wis., 1984, in press.

617. Colesante, R. T., Engstrom-Heg, R., Ehlinger, N., and Youmans, N., Cause and control of muskellunge fry mortality at Chautauqua hatchery, New York, *Prog. Fish-Cult.*, 43, 17, 1981.

618. Schachte, J. H., Iodophor disinfection of muskellunge eggs under intensive culture in hatcheries, *Prog. Fish-Cult.*, 41, 189, 1979.

619. Boersen, G. L. and Westers, H., Solids control in a baffled raceway, submitted for publication.

620. Westers, H., Principles of Intensive Fish Culture, Michigan Department of Natural Resources, 1981, 1.

621. Westers, H. and Copeland, J., The art of intensive esocid culture in Michigan, 1983, 1.

622. Svetovidov, A. N. and Dorofeeva, E. A., Systematics, origin, and history of the distribution of the Eurasian and North American perches and pike-perches (genera *Perca, Lucioperca,* and *Stizostedion*), *Vop. Ikhtiol.*, 3, 625, 1963.

623. Scott, W. B. and Crossman, E. J., Provisional Checklist of Canadian Freshwater Fishes, Information Leaflet, Department Ichthyology and Herpetology, Royal Ontario Museum, July 1967, 1.

624. McPhail, J. D. and Lindsey, C. C., Freshwater fishes of northwestern Canada and Alaska, *Fish. Res. Bd. Can. Bull.*, 173, 1970, 1.

625. Collette, B. B. and Banarescu, P., Systematics and zoogeography of the family Percidae, *J. Fish. Res. Bd. Can.*, 34, 1450, 1977.

626. Thorpe, J. E., Morphology, physiology, behavior, and ecology of *Perca fluviatilis* L. and *P. flavescens* Mitchill, *J. Fish. Res. Bd. Can.*, 34, 1504, 1977.

627. Scott, W. B. and Crossman, E. J., Freshwater fishes of Canada, *Fish. Res. Bd. Can. Bull.*, 184, 1973, 1.

628. Leach, G. C., Propagation and Distribution of Food Fishes, Fiscal Year 1927, Report to the U.S. Commissioner of Fisheries, 1928, 683.

629. Muncy, R. J., Evaluation of the Yellow Perch Hatchery Program in Maryland, Resource Study Rep., Maryland Department Resources Education, 15, 1959, 1.

630. Muncy, R. J., Life history of the yellow perch, *Perca flavescens*, in estuarine waters of Seven River, a tributary of Chesapeake Bay, Maryland, *Ches. Sci.*, 6, 545, 1962.

631. Follett, R., Raising Perch for the Midwest Market, Advisory Report 13, University of Wisconsin Sea Grant Program, Madison, 1975, 1.

632. Lesser, W. H. and Vilstrup, R., The Supply and Demand for Yellow Perch 1915—1990, University of Wisconsin College of Agriculture Life Science Res. Bull. R3006, 1979, 1.

633. Downs, W., Wisconsin: the dairy state takes a look at fish farming. Raising perch for the midwest market, *Comm. Fish. Farm.*, 1(5), 27, 1975.

634. Clugston, J. P., Oliver, J. L., and Ruelle, R., Reproduction, growth, and standing crops of yellow perch in southern reservoirs, in *Selected Coolwater Fishes of North America*, Special Publ. No. 11, Kendall, R. L., Ed., American Fisheries Society, Washington, D.C., 1978, 89.

635. Lagler, K. F., Bardach, J. E., and Miller, R. R., *Ichthyology*, John Wiley & Sons, New York, 1962, 1.

636. Hutchinson, B., Yellow Perch Egg and Prolarvae Mortality, Fish and Wildlife Restoration Project Report, 1974, 1.

637. Brazo, D. C., Tack, D. I., and Liston, C. R., Age, growth, and fecundity of yellow perch, *Perca flavescens*, in Lake Michigan, *Trans. Am. Fish. Soc.*, 104, 726, 1975.

638. West, G. and Leonard, J., Culture of yellow perch with emphasis on development of eggs and fry, in *Selected Coolwater Fishes of North America,* Special Publ. No. 11, Kendall, R. L., Ed., American Fisheries Society, Washington, D.C., 1978, 172.

639. Sheri, A. N., and Posner, G., Fecundity of the yellow perch, *Perca flavescens* (Mitchill), in the Bay of Quinte, Lake Ontario, *Can. J. Zool.,* 47, 55, 1969.

640. Herman, E., Wisly, W., Wiegert, L., and Burdick, M., The Yellow Perch: its Life History, Ecology, and Management, Wisconsin Conservation Department Publ. 228, 1959, 1.

641. Clady, M. D., Influence of temperature and wind on the survival of early stages of perch, *Perca flavescens, J. Fish. Res. Bd. Can.,* 33, 1887, 1976.

642. Harrington, R. W., Observations on the breeding habits of the yellow perch, *Perca flavescens* (Mitchill), *Copeia,* 1947, 199, 1947.

643. Hergenrader, G. L., Spawning behavior of *Perca flavescens* in aquaria, *Copeia,* 1969, 839, 1969.

644. Kayes, T., Reproductive biology and artificial propagation methods for adult perch, in Perch Fingerling Production for Aquaculture, Soderberg, R. W., Ed., Advisory Report 421, University of Wisconsin Sea Grant Program, Madison, 1977, 6.

645. Mansueti, A. J., Early development of the yellow perch, *Perca flavescens, Ches. Sci.,* 5, 46, 1964.

646. Worth, S. G., Observations on hatching of yellow perch, *Bull. U.S. Fish Comm. for 1890,* 10, 331, 1892.

647. Hokanson, K. E. F. and Kleiner, C. F., Effects of constant and rising temperatures on survival and developmental rates of embryonic and larval yellow perch, *Perca flavescens* (Mitchill), in *The Early Life History of Fish,* Blaxter, J.S., Ed., Springer-Verlag, New York, 1974, 437.

648. Hokanson, K. E. F., Temperature requirements of some percids and adaptations to the seasonal temperature cycle, *J. Fish. Res. Bd. Can.,* 34, 1524, 1977.

649. Kayes, T. B. and Calbert, H. E., Effects of photoperiod and temperature on the spawning of yellow perch (*Perca flavescens*), *Proc. World Maricult. Soc.,* 10, 306, 1979.

650. Koenig, S. T., Kayes, T. B., and Calbert, H. E., Preliminary observations on the sperm of yellow perch, in *Selected Coolwater Fishes of North America,* Special Publ. No. 11, Kendall, R. L., Ed., American Fisheries Society, Washington, D.C., 177, 1978.

651. Goetz, F. W. and Bergman, H. L., The *in vitro* effects of mammalian and piscine gonadotropin and pituitary preparations on final maturation in yellow perch (*Perca flavescens*) and walleye (*Stizostedion vitreum*), *Can. J. Zool.,* 56, 348, 1978.

652. Goetz, F. W. and Bergman, H. L., The effects of steroids on final maturation and ovulation of oocytes from brook trout (*Salvelinus fontinalis*) and yellow perch (*Perca flavescens*), *Biol. Reprod.,* 18, 293, 1978.

653. Wiggins, T. A., Bender, T. R., Mudrak, V. A., and Takacs, M. A., Hybridization of yellow perch and walleye, *Prog. Fish-Cult.,* 45, 131, 1983.

654. Ross, J., Powles, P. M., and Berrill, M., Thermal selection and related behavior in larval yellow perch (*Perca flavescens*), *Can. Field Nat.,* 91, 406, 1977.

655. Houde, E. D., Sustained swimming ability of larvae of walleye (*Stizostedion vitreum*) and yellow perch (*Perca flavescens*), *J. Fish. Res. Bd. Can.,* 26, 1647, 1969.

656. Hokanson, K. E. F., Optimum culture requirements of early life phases of yellow perch, in Perch Fingerling Production for Aquaculture, Soderberg, R. W., Ed., University of Wisconsin Sea Grant Advisory Rep. 421, Madison, 1977, 24.

657. McCauley, R. W. and Read, L. A. A., Temperature selection by juvenile and adult yellow perch (*Perca flavescens*) acclimated to 24 C, *J. Fish. Res. Bd. Can.,* 30, 1253, 1973.

658. Hale, J. G. and Carlson, A. R., Culture of the yellow perch in the laboratory, *Prog. Fish-Cult.,* 34, 195, 1972.

659. McCormick, J. H., Temperatures Suitable for the Well-Being of Juvenile Yellow Perch During their First Summer-Growing Season, Annual Report, ROAP-16AB1, National Water Quality Laboratory, Duluth, Minn., 1974, 1.

660. McCormick, J. H., Temperature Effects on Young Yellow Perch, *Perca flavescens* (Mitchill), EPA-600/3-76-057, U.S. Environmental Protection Agency Ecology Research Service, Washington, D.C., 1976, 1.

661. Huh, H. T., Bioenergetics of Food Conversion and Growth of Yellow Perch (*Perca flavescens*) and Walleye (*Stizostedion vitreum vitreum*) Using Formulated Diets, Ph.D. dissertation, University of Wisconsin, Madison, 1975, 1.

662. Stairs, S., Experience in perch fingerling production from the Lake Mills National Fish Hatchery, in Perch Fingerling Production for Aquaculture, Soderberg, R. W., Ed., Advisory Report 421, University of Wisconsin Sea Grant Program, Madison, 1977, 58.

663. Best, C. D., Initiation of Artificial Feeding and the Control of Sex Differentiation in Yellow Perch, *Perca flavescens,* M.S. thesis, University of Wisconsin, Madison, 1981, 1.

664. Carlander, K. D., *Handbook of Freshwater Fisheries Biology,* Wm. C. Brown Co., Dubuque, Iowa, 1950, 1.

665. Calbert, H. E. and Huh, H. T., Culturing yellow perch *Perca flavescens* under controlled environmental conditions for the upper midwest market, *Proc. World Maricult. Soc.*, 7, 137, 1976.
666. Huh, H. T., Calbert, H. E., and Stuiber, D. A., Effects of temperature and light on growth of yellow perch and walleye using formulated feed, *Trans. Am. Fish. Soc.*, 105, 254, 1976.
667. Carlson, A. R., Blocker, J., and Herman, L. J., Growth and survival of channel catfish and yellow perch exposed to lowered constant and diurnally fluctuating dissolved oxygen concentrations, *Prog. Fish-Cult.*, 42, 73, 1980.
668. Schott, E. F., Kayes, T. B., and Calbert, H. E., Comparative growth of male versus female yellow perch fingerlings under controlled environmental conditions, in *Selected Coolwater Fishes of North America,* Special Publ. No. 11, Kendall, R. L., Ed., American Fisheries Society, Washington, D.C., 1978, 181.
669. Kayes, T. B., Beat, C. D., and Malison, J. A., Hormonal manipulation of sex in yellow perch (*Perca flavescens),* Am. Zool., 22, 955, 1983.
670. Schumann, G. O., Artificial light to attract young perch: a new method of augmenting the food supply of predacious fish fry in hatcheries, *Prog. Fish-Cult.*, 25, 171, 1963.
671. Manci, W. E., Malison, J. A., Kayes, T. B., and Kuczynaki, T. E., Harvesting photopositive juvenile fish from a pond using a lift net and light, *Aquaculture,* 34, 157, 1983.
672. Lewis, W. M., Heidinger, R. C., and Tetzlaff, B. L., Tank Culture of Striped Bass, Fisheries Research Laboratory, Southern Illinois University, Carbondale, 1981, 1.
673. Reinitz, G. and Austin, R., Experimental diets for intensive culture of yellow perch, *Prog. Fish-Cult.,* 42, 29, 1980.
674. Kitchell, J. F., Steward, D. J., and Weininger, D., Applications of a bioenergetics model to yellow perch (*Perca flavescens)*and walleye (*Stizostedion vitreum vitreum), J. Fish. Res. Bd. Can.,* 34, 1922, 1977.
675. Kocurek, D., An Economic Study of a Recirculating Perch Aquaculture System, M.S. thesis, University of Wisconsin, Madison, 1979, 1.
676. Lesser, W. H., Marketing Systems for Warm Water Aquaculture Species in the Upper Midwest, Ph.D. dissertation, University of Wisconsin, Madison, 1978, 1.
677. Colby, P. J., McNicol, R. E., and Ryder, R. A., Synopsis on Biological Data on the Walleye *Stizostedion v. vitreum* (Mitchill 1818), FAO Fish. Synop. 119, Food and Agriculture Organization, Rome, 1979, 1.
678. Baldwin, N. S. and Sealfeld, R. W., Commercial Fish Production in the Great Lakes, 1867—1960, Great Lakes Fishery Commission, Ann Arbor, Mich., 1962, 1.
679. Webster, J., et al., Historical Perspective of Propagation and Management of Coolwater Fishes in the United States, Spec. Publ. No. 11, American Fisheries Society, Washington, D.C., 1978, 161.
680. Cobb, E. W., Pike-perch propagation in northern Minnesota, *Trans. Am. Fish. Soc.,* 53, 95, 1923.
681. Nevin, J., Hatching the wall-eyed pike, *Trans. Am. Fish. Soc.,* 16, 14, 1887.
682. Nickum, J. G., Intensive culture of walleyes: the state of the art, in *Selected Coolwater Fishes of North America,* Special Publ. No. 11, Kendall, R. L., Ed., American Fisheries Society, Washington, D.C., 1978, 187.
683. Johnson, F. H., Environmental and species associations of the walleye in Lake Winnibigoshish and connected waters, including observations on food habits and predator-prey relationships, *Minn. Fish Invest.,* 5, 5, 1969.
684. Kelso, J. R. M., Diel movement of walleye, *Stizostedion vitreum vitreum,* in West Blue Lake, Manitoba, as determined by ultrasonic tracking. *J. Fish. Res. Bd. Can.,* 33, 2070, 1976.
685. Ferguson, R. G., The preferred temperature of fish and their midsummer distribution in temperate lakes and streams, *J. Fish. Res. Bd. Can.,* 15, 607, 1958.
686. Scherer, E., Overhead-light intensity and vertical positioning of the walleye, *Stizostedion vitreum vitreum, J. Fish. Res. Bd. Can.,* 33, 289, 1976.
687. Ryder, R. A., Effects of ambient light variations on behavior of yearling, subadult, and adult walleyes (*Stizostedion vitreum vitreum), J. Fish. Res. Bd. Can.,* 34, 1481, 1977.
688. Ali, M. A. and Anctil, M., Corrélation entre la structure rétinienne et l'habitat chez *Stizostedion vitreum vitreum* et *S. canadense, J. Fish. Res. Bd. Can.,* 25, 2001, 1968.
689. Scherer, E., Effects of oxygen depletion and of carbon dioxide buildup on the photic behavior of the walleye (*Stizostedion vitreum vitreum), J.Fish. Res. Bd. Can.,* 28, 1303, 1971.
690. Dendy, J. S., Predicting depth distribution of fish in three TVA storage type reservoirs, *Trans. Am. Fish. Soc.,* 75, 65, 1948.
691. Eschmeyer, P. H., The life history of the walleye, *Stizostedion vitreum vitreum* (Mitchill), in Michigan, *Bull. Mich. Dept. Conserv., Inst. Fish. Res.,* 3, 1, 1950.
692. Ellis, D. V. and Giles, M. A., The spawning behaviour of the walleye, *Stizostedion vitreum* (Mitchill), *Trans. Am. Fish. Soc.,* 94, 358, 1965.
693. Priegel, G. R., Reproduction and Early Life History of the Walleye in the Lake Winnebago Region, Tech. Bull. 45, Wisconsin Department Natural Resources, Madison, 1970, 1.

694. Smith, C. G., Egg production of walleyed pike and sauger, *Prog. Fish-Cult.,* 54, 32, 1941.
695. Arnold, B. B., Life History Notes on the Walleye, *Stizostedion vitreum vitreum* in a Turbid Water Utah Lake, M.S. thesis, Utah State University, Logan, 1960, 1.
696. Wolfert, D. R., Maturity and fecundity of walleyes from the eastern and western basins of Lake Erie, *J. Fish. Res. Bd. Can.,* 26, 1877, 1969.
697. Muench, K. A., Certain Aspects of the Life History of the Walleye, *Stizostedion vitreum vitreum* in Center Hill Reservoir, Tennessee, M.S. thesis, Tennessee Technical University Cookeville, 1966, 1.
698. Cook, F. A., Freshwater Fishes of Mississippi, Mississippi Game and Fish Commission, Jackson, 1959, 1.
699. Niemuth, W., Churchill, W., and Wirth, T., The walleye, its life history, ecology and management, *Publ. Wisc. Conserv. Dept.,* 227, 1, 1966.
700. Nelson, W. R., Hines, N. R., and Beckman, L. G., Artificial propagation of saugers and hybridization with walleyes, *Prog. Fish-Cult.,* 27, 216, 1965.
701. Koenst, W. M. and Smith, L. L., Jr., Thermal requirements of the early life history stages of walleye, *Stizostedion vitreum vitreum* and sauger, *Stizostedion canadense, J. Fish. Res. Bd. Can.,* 33, 1130, 1976.
702. Baker, C. T. and Scholl, R. L., Walleye Spawning Area Study in Western Lake Erie, Fish and Wildlife Restoration Project Report, unpublished, 1969, 1.
703. Johnson, F. H., Walleye egg survival during incubation on several types of bottom in Lake Winnibigoshish, Minnesota, and connecting waters, *Trans. Am. Fish. Soc.,* 90, 312, 1961.
704. Hohn, M. H., Analysis of plankton ingested by *Stizostedium* (sic) *virtreum vitreum* (Mitchill) fry and concurrent vertical plankton tows from southwestern Lake Erie, May 1961 and May 1962, *Ohio J. Sci.,* 66, 193, 1966.
705. Paulus, R. D., Walleye Fry Food Habits in Lake Erie, Ohio Fish Monogr., Ohio Department Natural Resources, Division Wildlife Report, 1972, 1.
706. Raisenen, G. A. and Applegate, R. L., Prey selection of walleye fry in an experimental system, *Prog. Fish-Cult.,* 45, 209, 1983.
707. Bandow, J., Methods for increasing the growth rate of walleye fingerlings in ponds, *Minn. Dept. Nat. Resour. Performance Rep.,* 302, 1, 1975.
708. Cobb, E. W., Pike-Perch propagation in northern Minnesota, *Trans. Am. Fish Soc.,* 53, 95, 1923.
709. Lessman, C. A., Effects of gonadotropin mixtures and two steroids on inducing ovulation in the walleye, *Prog. Fish-Cult.,* 40, 3, 1978.
710. Hearn, M. C., Ovulation of pond-reared walleyes in response to various injection levels of human chorionic gonadotropin, *Prog. Fish-Cult.,* 42, 228, 1980.
711. Heidinger, R. C., personal communication, 1985.
712. Olson, D. E., Improvement of Artificial Fertilization Methods at a Walleye Hatchery, Invest. Rep. Minn. DNR (310), 1971, 1.
713. Waltemyer, D. L., The effect of tannin on the motility of walleye (*Stizostedion vitreum*) spermatozoa. *Trans. Am. Fish. Soc.,* 104, 808, 1975.
714. Woynarovich, E., Über die kunstliche vermehrung des karpfens und erbrutung des laiches in zugerglasern, *Wasser Abwasser. Beitr. Gewasserforsch.,* 4, 210, 1964.
715. Rubstov, V. V., The effect of wild carp (*Cyprinus carpio*) egg membranes of de-gumming with PAS-G and tannin, *J. Ichthyol.,* 16, 1028, 1976.
716. Dumas, R. F. and Brand, J. S., Use of tannin solution in walleye and carp culture, *Prog. Fish-Cult.,* 34, 7, 1972.
717. Colesante, R. T. and Youmans, N. B., Water-hardening walleye eggs with tannic acid in a production hatchery, *Prog. Fish-Cult.,* 45, 126, 1983.
718. Jahncke, M. L., Selected Factors Influencing Mortality of Walleye Fry in Intensive Culture, M.S. thesis, Cornell University, Ithaca, N.Y., 1981, 1.
719. Houde, E. D., Food of pelagic young of the walleye, *Stizostedion vitreum vitreum* in Oneida Lake, New York, *Trans. Am. Fish. Soc.,* 96, 17, 1967.
720. Smith, L. L., Jr. and Moyle, J. B., Factors influencing production of yellow pike-perch *Stizostedion vitreum vitreum* in Minnesota rearing ponds, *Trans. Am. Fish. Soc.,* 73, 243, 1945.
721. Dobie, J. and Moyle, J. B., Methods Used for Investigating Productivity of Fish Rearing Ponds in Minnesota, Spec. Publ. 5, Minnesota Department Conservation, 1956.
722. Hohn, M. H., Analysis of plankton ingested by *Stizostedion vitreum vitreum* fry and concurrent vertical plankton tows from southwestern Lake Erie, May, 1961 and May, 1962, *Ohio J. Sci.,* 66, 193, 1966.
723. Paulus, R. D., Walleyefry Food Habits in Lake Erie, Ohio Fish Monogr. 2, Ohio Department Natural Resources, 1969, 1.
724. Walker, R. E. and Applegate, R. L., Growth, food, and possible ecological effects of young-of-the-year walleyes in a South Dakota prairie pothole, *Prog. Fish-Cult.,* 38, 217, 1976.

725. Dobie, J., Minnesota walleye nursery ponds and transportation of fingerling walleye, in Proc. North Central Warmwater Fish Culture-Management Workshop, Iowa Cooperative Fishery Research Unit, Ames, 1971, 133.
726. Geiger, J. G., Zooplankton production and manipulation in striped bass rearing ponds, *Aquaculture,* 35, 331, 1983.
727. Cheshire, W. F. and Steele, K. L., hatchery rearing of walleyes using artificial food, *Prog. Fish-Cult.,* 34, 96, 1972.
728. National Task Force for Public Fish Hatchery Policy, Report of the National Task Force for Public Fish Hatchery Policy, U.S. Fish and Wildlife Service, Washington, D.C., 1974, 1.
729. McCauley, R. W., Automatic food pellet dispenser for walleyes, *Prog. Fish-Cult.,* 32, 42, 1970.
730. Graves, G., They said it couldn't be done, *Farm Pond Harvest,* 8(1), 6ff, 1974.
731. Nagel, T. O., Rearing of walleye fingerlings in an intensive culture using Oregon Moist Pellets as an artificial diet, *Prog. Fish-Cult.,* 36, 59, 1974.
732. Nagel, T. O., Intensive culture of fingerling walleyes on formulated feeds, *Prog. Fish-Cult.,* 38, 90, 1976.
733. Beyerle, G. B., Summary of attempts to raise walleye fry and fingerlings on artificial diets, with suggestions on needed research, and procedures to be used in future tests, *Prog. Fish-Cult.,* 37, 103, 1975.
734. Corazza, L. and Nickum, J. G., Rate of food passage through the gastrointestinal tract of fingerling walleye, *Prog. Fish-Cult.,* 45, 183, 1983.
735. Kostomarova, A. A., Significance of the phase of mixed feeding for the survival of pike larvae, in Proc. Conf. Population Dynamics of Fishes, Ministry of Agriculture, Fisheries, and Food, English Translation prepared on behalf of Fisheries Laboratory, Lowestoft, Suffolk, England, 1961, 344.
736. Cuff, W. R., Initiation and control of cannibalism in larval walleyes, *Prog. Fish-Cult.,* 39, 29, 1977.
737. Mathias, J. H. and Li, S., Feeding habits of walleye larvae and juveniles: comparative laboratory and field studies, *Trans. Am. Fish. Soc.,* 111, 722, 1982.
738. Li, S. and Mathias, J. A., Causes of high mortality among cultured larval walleyes, *Trans. Am. Fish. Soc.,* 111, 710, 1982.
739. Raisanen, G. A. and Applegate, Prey selection of walleye fry in an experimental system, *Prog. Fish-Cult.,* 45, 209, 1983.
740. Hnath, J. G., A summary of fish diseases and treatments administered in a coolwater diet testing program, *Prog. Fish-Cult,* 37, 106, 1975.
741. Davis, E. M., unpublished information.
742. Raney, E. C., The life history of the striped bass, *Roccus saxatilis* (Walbaum), *Bull. Bingham Ocean. Coll.,* 14, 5, 1952.
743. Barkaloo, J. M., Florida striped bass, *Florida Game Freshw. Comm. Fish. Bull.,* 4, 1, 1967.
744. McIlwain, T. D., Distribution of the striped bass, *Roccus saxatilis* (Walbaum), in Mississippi waters, *Proc. S.E. Assoc. Game Fish Comm.,* 21, 254, 1968.
745. Radovich, J., Effect of ocean temperature on the seaward movement of striped bass, *Roccus saxatilis,* on the Pacific coast, *Calif. Fish Game,* 49, 191, 1963.
746. Bailey, W. M., An evaluation of striped bass introductions in the southeastern United States, *Proc. S.E. Assoc. Game Fish Comm.,* 28, 54, 1975.
747. Stevens, R. E., Historical overview of striped bass culture and management, in The Aquaculture of Striped Bass: A Proceedings, McCraren, J. P., Ed., University of Maryland Sea Grant Publ. UM-SG-MAP-84-01, College Park, 1984, 1.
748. Talbot, G. B., Estuarine Environmental Requirements and Limiting Factors for Striped Bass, American Fisheries Society Special Publication No. 3, 1966, 3.
749. Forrester, C. R., Peden, A. E., and Wilson, R. M., First records of the striped bass, *Morone saxatilis,* in British Columbia waters, *J. Fish. Res. Bd. Can.,* 29, 337, 1972.
750. Bishop, R. D., Evaluation of the striped bass (*Roccus saxatilis*) and white bass (*R. chrysops*) hybrids after two years, *Proc. S.E. Assoc. Game Fish Comm.,* 21, 245, 1968.
751. Smith, W. B., Bonner, W. R., and Tatum, B. L., Premature egg procurement from striped bass, *Proc. S.E. Assoc. Game Fish Comm.,* 20, 324, 1967.
752. Bayless, J. D., Striped bass hatching and hybridization experiments, *Proc. S.E. Assoc. Game Fish Comm.,* 21, 233, 1968.
753. Kerby, J. H., Feasibility of Artificial Propagation and Introduction of Hybrids of the *Morone* Complex into Estuarine Environments, with a Meristic and Morphometric Description of the Hybrids, Ph.D. dissertation, University of Virginia, Charlottesville, 1972, 1.
754. Ware, F. J., Progress with *Morone* hybrids in fresh water, *Proc. S.E. Assoc. Game Fish Comm.,* 28, 48, 1975.
755. Logan, H. J., Comparison of growth and survival rates of striped bass and striped bass X white bass hybrids under controlled environments, *Proc. S.E. Assoc. Game Fish Comm.,* 21, 260, 1968.

756. Williams, H. M., Preliminary studies of certain aspects of the life history of the hybrid (striped bass x white bass) in two South Carolina reservoirs, *Proc. S.E. Assoc. Game Fish Comm.*, 24, 424, 1971.

757. Bonn, E. W., Bailey, W. M., Bayless, J. D., Erickson, K. E., and Stevens, R. E., Eds., *Guidelines for Striped Bass Culture,* Striped Bass Committee, Southern Division, American Fisheries Society, 1976, 1.

758. Kerby, J. H. and Joseph, E. B., Growth and survival of striped bass and striped bass x white bass hybrids, *Proc. S.E. Assoc. Game Fish Comm.*, 32, 715, 1979.

759. Striped Bass Committee, Southern Division, American Fisheries Society, 1982.

760. Jordan, D. S. and Evermann, B. W., *American Food and Game Fishes,* Doubleday, Page and Company, New York, 1902, 1.

761. Worth, S. G., Report upon the propagation of striped bass at Weldon, N. C., in the spring of 1884, *Bull. U.S. Fish Comm.*, 4, 225, 1884.

762. Worth, S. G., The artificial propagation of the striped bass (*Roccus lineatus*) on Albemarle Sound, *Bull. U.S. Fish Comm.*, 1, 174, 1882.

763. Tatum, B. L., Bayless, J. D., McCoy, E. G., and Smith, W. B., Preliminary experiments in the artificial propagation of striped bass, *Roccus saxatilis, Proc. S.E. Assoc. Game Fish Comm.*, 19, 374, 1966.

764. Scruggs, G. D., Jr., Reproduction of resident striped bass in Santee-Cooper reservoir, South Carolina, *Trans. Am. Fish. Soc.*, 85, 144, 1957.

765. Stevens, R. E., The striped bass of the Santee-Cooper reservoir, *Proc. S.E. Assoc. Game Fish Comm.*, 11, 253, 1958.

766. Fuller, J. C., Jr., South Carolina's Striped Bass Story, South Carolina Wildlife Resources Department, 1968, 1.

767. Surber, E. W., Results of striped bass (*Roccus saxatilis*) introductions into freshwater impoundments, *Proc. S.E. Assoc. Game Fish Comm.*, 11, 273, 1958.

768. Stevens, R. E., A final report on the use of hormones to ovulate striped bass, *Roccus saxatilis* (Walbaum), *Proc. S.E. Assoc. Game Fish Comm.*, 18, 525, 1967.

769. Sandoz, O. and Johnston, K. H., Culture of striped bass, *Proc. S.E. Assoc. Game Fish Comm.*, 19, 390, 1966.

770. Stevens, R. E., May, O. D., Jr., and Logan, H. J., An interim report on the use of hormones to ovulate striped bass (*Roccus saxatilis*), *Proc. S.E. Assoc. Game Fish Comm.*, 17, 226, 1965.

771. Stevens, R. E., Hormone-induced spawning of striped bass for reservoir stocking, *Prog. Fish-Cult.*, 28, 19, 1966.

772. Bayless, J. D., Artificial Propagation and Hybridization of Striped Bass, *Morone saxatilis* (Walbaum), South Carolina Wildlife and Marine Resources Department, 1972, 1.

773. Bishop, R. D., The use of circular tanks for spawning striped bass (*Morone saxatilis*), *Proc. S.E. Assoc. Game Fish Comm.*, 18, 35, 1968.

774. McGill, E. M., Jr., Pond water for rearing striped bass fry, *Roccus saxatilis* (Walbaum), in aquaria, *Proc. S.E. Assoc. Game Fish Comm.*, 20, 331, 1967.

775. Anderson, J. C., Production of striped bass fingerlings, *Prog. Fish-Cult.*, 28, 162, 1966.

776. Regan, D. M., Wellborn, T. L., and Bowker, R. G., Striped bass, *Roccus saxatilis* (Walbaum), Development of Essential Requirements for Production, U.S. Fish and Wildlife Service, Atlanta, Ga., 1968, 1.

777. Ray, R. H. and Wirtanen, L. J., Striped bass, *Morone saxatilis* (Walbaum), 1969 Report on the Development of Essential Requirements for Production, U.S. Fish and Wildlife Service, Atlanta, Ga., 1970, 1.

778. Wirtanen, L. J. and Ray, R. H., Striped bass, *Morone saxatilis* (Walbaum), 1970 Report on the Development of Essential Requirements for Production, U.S. Fish and Wildlife Service, Atlanta, Ga., 1971, 1.

779. Reeves, W. C. and Germann, J. F., Effects of increased water hardness, source of fry and age at stocking on survival of striped bass fry in earthen ponds, *Proc. S.E. Assoc. Game Fish Comm.*, 25, 542, 1972.

780. Harper, J. L. and Jarman, R., Investigation of striped bass, *Morone saxatilis* (Walbaum), culture in Oklahoma, *Proc. S.E. Assoc. Game Fish Comm.*, 25, 501, 1972.

781. Barwick, D. H., The effect of increased sodium chloride on striped bass fry survival in freshwater ponds, *Proc. S.E. Assoc. Game Fish Comm.*, 27, 415, 1974.

782. Hughes, J. S., Striped bass, *Morone saxatilis* (Walbaum), Culture Investigations in Louisiana with Notes of Sensitivity of Fry and Fingerlings to Various Chemicals, Fisheries Bull. 113, Louisiana Wildlife and Fisheries Commission, 1975, 1.

783. Bayless, J. D., personal communication, 1980.

784. Brashler, E. W., Development of pond culture techniques for striped bass *Morone saxatilis* (Walbaum), *Proc. S.E. Assoc. Game Fish Comm.*, 28, 44, 1975.

785. Harper, J. L., Jarman, R., and Yacovino, J. T., Food habits of young striped bass, *Roccus saxatilis* (Walbaum), in culture ponds, *Proc. S.E. Assoc. Game Fish Comm.*, 22, 373, 1969.
786. Meshaw, J. C., Jr., A Study of Feeding Selectivity of Striped Bass Fry and Fingerlings in Relation to Plankton Availability, M.S. thesis, North Carolina State University, Raleigh, 1969, 1.
787. Humphries, E. T. and Cumming, K. B., Food habits and feeding selectivity of striped bass fingerlings in culture ponds, *Proc. S.E. Assoc. Game Fish Comm.*, 25, 522, 1972.
788. Harrell, R. M., Loyacano, H. A., Jr., and Bayless, J. D., Zooplankton availability and selectivity of fingerling striped bass, *Georgia J. Sci.*, 35, 129, 1977.
789. Stevens, R. E., Current and future considerations concerning striped bass culture and management, *Proc. S.E. Assoc. Game Fish Comm.*, 28, 69, 1975.
790. Inslee, T. D., Holding striped bass larvae in cages until swim-up, *Proc. S.E. Assoc. Fish Wildl. Agen.*, 31, 422, 1977.
791. Striped Bass Committee, Southern Division, American Fisheries Society, personal communication, 1984.
792. Parker, N. C., Striped bass culture in continuously aerated ponds, *Proc. S.E. Assoc. Fish Wildl. Agen.*, 33, 353, 1980.
793. Rees, R. A. and Cook, S. F., Evaluation of optimum stocking rate of striped bass X white bass fry in hatchery rearing ponds, *Proc. S.E. Assoc. Fish Wildl. Agen.*, in press.
794. Humphries, E. T., and Cumming, K. B., An evaluation of striped bass fingerling culture, *Trans. Am. Fish. Soc.*, 102, 13, 1973.
795. Rhodes, W. and Merriner, J. V., A preliminary report on closed system rearing of striped bass sac fry to fingerling size, *Prog. Fish-Cult.*, 35, 199, 1973.
796. McIlwain, T. D., Closed recirculating system for striped bass production, *Proc. World Maricult. Soc.*, 7, 523, 1976.
797. Texas Instruments Incorporated, Feasibility of culturing and stocking Hudson River striped bass, an overview 1973 to 1975, Texas Instruments Incorporated, Ecological Services, Dallas, Tex., 1977, 1.
798. Nicholson, L. C., Culture of striped bass (*Morone saxatilis*) in raceways under controlled conditions, presented at *Western Assoc. State Game Fish Comm.*, 1973.
799. Nicholson, L. C., personal communication, 1982.
800. Powell, M. R., Cage and raceway culture of striped bass in brackish water in Alabama, *Proc. S.E. Assoc. Game Fish Comm.*, 26, 345, 1973.
801. Gomez, R., Food habits of young-of-the-year striped bass, *Roccus saxatilis* (Walbaum), in Canton Reservoir, *Proc. Okla. Acad. Sci.*, 50, 79, 1970.
802. Ager, L. M., Food Habits of Young-of-the-Year Striped Bass in Lake Sinclair, Project No. WC-4, Georgia Department of Natural Resources, Game and Fish Division, 1979, 1.
803. Heubach, W., Toth, R. J., and McCready, A. M., Food of young-of-the-year striped bass (*Roccus saxatilis*) in the Sacramento-San Joaquin River system, *Calif. Fish Game*, 49, 224, 1963.
804. Markle, D. F. and Grant, G. C., The summer food habits of young-of-the-year striped bass in three Virginia rivers, *Ches. Sci.*, 11, 50, 1970.
805. Kelley, J. R., Jr., Preliminary report on methods for rearing striped bass, *Roccus saxatilis* (Walbaum), fingerlings, *Proc. S.E. Assoc. Game Fish Comm.*, 20, 341, 1967.
806. Bowman, J. R., Survival and Growth of Striped Bass, *Morone saxatilis* (Walbaum), Fry Fed Artificial Diets in a Closed Recirculation System, Ph.D. dissertation, Auburn University, Auburn, Ala., 1979, 1.
807. Falls, W. W., Food Habits and Feeding Selectivity of Larvae of the Striped Bass *Morone saxatilis* (Walbaum) (Osteichthys; Percichthidae) Under Intensive Culture Conditions, Ph.D. dissertation, University of Southern Mississippi, Hattiesburg, 1983, 1.
808. Eldridge, M. B., King, D. J., Eng, D., and Bowers, M. J., Role of the oil globule in survival and growth of striped bass (*Morone saxatilis*) larvae, *Proc. West. Assoc. Game Fish Comm.*, 57, 303, 1977.
809. Eldridge, M. B., Whipple, J. A., Eng, D., Bowers, M. J., and Jarvis, B. M., Effects of food and feeding factors on laboratory-reared striped bass larvae, *Trans. Am. Fish. Soc.*, 110, 111, 1981.
810. Rogers, B. A. and Westin, D. T., Laboratory studies on effects of temperature and delayed initial feeding on development of striped bass larvae, *Trans. Am. Fish. Soc.*, 110, 100, 1981.
811. Daniel, D. A., A Laboratory Study to Define the Relationship Between Survival of Young Striped Bass *(Morone saxatilis)* and Their Food Supply, Administrative Report No. 76-1, California Department of Fish and Game, 1976, 1.
812. McHugh, J. J. and Heidinger, R. C., Effects of light on feeding and egestion time of striped bass fry, *Prog. Fish-Cult.*, 39, 33, 1977.
813. Valenti, R. J., Aldred, J., and Liebell, J., Experimental marine cage culture of striped bass in northern waters, *Proc. World Maricult. Soc.*, 7, 99, 1976.
814. Millikin, M. R., Effects of dietary protein concentration on growth, feed efficiency, and body composition of age-0 striped bass, *Trans. Am. Fish. Soc.*, 111, 373, 1982.

815. Millikin, M. R., Interactive effects of dietary protein and lipid on growth and protein utilization of age-0 striped bass, *Trans. Am. Fish. Soc.,* 112, 185, 1983.

816. Smith, R. E. and Kernehan, R. J., Predation by the free-living copepod *Cyclops bicuspidatus thomasi,* on larvae of the striped bass and white perch, *Estuaries,* 4, 81, 1981.

817. Collins, C. M., Burton, G. L., and Schweinforth, R. L., High Density Culture of White Bass X Striped Bass Fingerlings in Raceways Using Power Plant Effluent, Tech. Rep. Series TVA/ONR/WR-83-11, Division Air and Water Resources, Office of Natural Resources, Tennessee Valley Authority, 1983, 1.

818. Braid, M. R. and Shell, E. W., Incidence of cannibalism among striped bass fry in an intensive culture system, *Prog. Fish-Cult.,* 43, 210, 1981.

819. U.S. Dept. Commerce, National Marine Fisheries Serv., Market News Rep., 1981—1982.

820. Swartz, D., Marketing striped bass, in The Aquaculture of Striped Bass: A Proceedings, McCraren, J. P., Ed., University of Maryland Sea Grant Publ. UM-SG-MAP-84-01, College Park, 1984, 233.

821. Stevens, R. E., Striped bass culture in the United States, *Comm. Fish. Farmer,* 5(3), 10, 1979.

822. Williams, J. E., Sandifer, P. A., and Lindbergh, J. M., Net-pen culture of striped bass X white bass hybrids in estuarine waters of South Carolina: a pilot study, *J. World Maricult. Soc.,* 12(2), 98, 1981.

823. Wawronowicz, L. J. and Lewis, W. M., Evaluation of the striped bass as a pond-reared food fish, *Prog. Fish-Cult.,* 41, 138, 1979.

824. Kerby, J. H., Woods, L. C., III, and Huish, M. T., Pond culture of striped bass X white bass hybrids, *J. World Maricult. Soc.,* 14, 613, 1983.

825. Kerby, J. H., Woods, L. C., III, and Huish, M. T., Culture of the striped bass and its hybrids: a review of methods, advances, and problems, in *Proc. of the Warmwater Fish Culture Workshop,* Stickney, R. R. and Meyers, S. P., Eds., Spec. Publ. No. 3, World Mariculture Society, 1984, 23.

826. Woods, L. C., III, Kerby, J. H., and Huish, M. T., Estuarine cage culture of hybrid striped bass, *J. World Maricult. Soc.,* 14, 595, 1983.

827. Woods, L. C., III, Kerby, J. H., and Huish, M. T., Culture of hybrid striped bass to marketable size in circular tanks, *Prog. Fish-Cult.,* 47, 147, 1985.

828. Shannon, E. H., Effect of temperature changes upon developing striped bass eggs and fry, *Proc. S.E. Assoc. Game Fish Comm.,* 23, 264, 1970.

829. Albrecht, A. B., Some observations on factors associated with survival of striped bass eggs and larvae, *Calif. Fish Game,* 50, 100, 1964.

830. Doroshev, S. I., Biological features of the eggs, larvae and young of the striped bass [*Roccus saxatilis* (Walbaum)] in connection with the problem of acclimatization in the USSR, *J. Ichthyol.,* 10, 235, 1970.

831. Otwell, W. S. and Merriner, J. V., Survival and growth of juvenile striped bass, *Morone saxatilis,* in a factorial experiment with temperature, salinity and age, *Trans. Am. Fish. Soc.,* 104, 560, 1975.

832. Morgan, R. P., II and Rasin, V. J., Jr., Temperature and salinity effects on development of striped bass eggs and larvae, *Trans. Am. Fish. Soc.,* 110, 95, 1981.

833. Shannon, E. H. and Smith, W. B., Preliminary observations on the effect of temperature on striped bass eggs and sac fry, *Proc. S.E. Assoc. Game Fish Comm.,* 21, 257, 1968.

834. Davies, W. D., Rates of temperature acclimation for hatchery reared striped bass fry and fingerlings, *Prog. Fish-Cult.,* 35, 214, 1973.

835. Cox, D. K. and Coutant, C. C., Growth dynamics of juvenile striped bass as functions of temperature and ration, *Trans. Am. Fish. Soc.,* 110, 226, 1981.

836. Woiwode, J. G. and Adelman, I. R., Growth, food conversion efficiency, and survival of the hybrid white X striped bass as a function of temperature, in The Aquaculture of Striped Bass: A Proceedings, McCraren, J. P., Ed., University of Maryland Sea Grant Publ. UM-SG-MAP-84-01, College Park, 1984, 143.

837. Turner, J. L. and Farley, T. C., Effects of temperature, salinity, and dissolved oxygen on the survival of striped bass eggs and larvae, *Calif. Fish Game,* 57, 268, 1971.

838. Harrell, R. M. and Bayless, J. D., Effects of suboptimal dissolved oxygen concentrations on developing striped bass embryos, *Proc. S.E. Assoc. Fish Wildl. Agen.,* 35, 508, 1981.

839. Hill, L. G., Schnell, G. D., and Matthews, W. J., Locomotor responses of the striped bass, *Morone saxatilis,* to environmental variables, *Am. Midl. Nat.,* 105, 139, 1981.

840. Chittenden, M. E., Jr., Effects of handling and salinity on oxygen requirements of the striped bass, *Morone saxatilis, J. Fish. Res. Bd. Can.,* 28, 1823, 1971.

841. Parker, N. C., Culture requirements for striped bass, in The Aquaculture of Striped Bass: A Proceedings, McCraren, J. P., Ed., University of Maryland Sea Grant Publ. UM-SG-MAP-84-01, College Park, 1984, 29.

842. Klyashtorin, L. B. and Yarzhombek, A. A., Some aspects of the physiology of the striped bass, *Morone saxatilis, J. Ichthyol.,* 15, 985, 1975.

843. Lal, K., Lasker, R., and Klujis, A., Acclimation and rearing of striped bass larvae in sea water, *Calif. Fish Game,* 63, 210, 1977.

844. Van Olst, J. C., Carlberg, J. M., Massingill, M. J., Hovanec, T. A., Cochran, M. D., and Doroshev, S. I., Methods for Intensive Culture of Striped Bass Larvae at Central Valleys Hatchery, Progress Report, California Fish and Game, University of California at Davis and Aquaculture Systems International, Elk Grove, Calif., 1980, 1.

845. Wattendorf, R. J. and Shafland, P. L., Observations on salinity tolerance of striped bass X white bass hybrids in aquaria, *Prog. Fish-Cult.*, 44, 148, 1982.

846. Hazel, C. R., Thomsen, W., and Meith, S. J., Sensitivity of striped bass and stickleback to ammonia in relation to temperature and salinity, *Calif. Fish Game*, 57, 154, 1971.

847. Siddal, S. E., Studies of closed marine culture systems, *Prog. Fish-Cult.*, 36, 8, 1974.

848. Davies, W. D., The effects of total dissolved solids, temperature, and pH on the survival of immature striped bass: a response surface experiment, *Prog. Fish-Cult.*, 35, 157, 1973.

849. Auld, A. H. and Schubel, J. R., Effects of suspended sediment on fish eggs and larvae: a laboratory assessment, *Est. Coast. Mar. Sci.*, 6, 153, 1977.

850. Setzler, E. M., Boynton, W. R., Wood, K. V., Zion, H. H., Lubbers, L., Moutford, N. K., Frere, P., Tucker, L., and Mihursky, J. A., Synopsis of Biological Data on Striped Bass, *Morone saxatilis* (Walbaum), NOAA Technical Report, NMFS Circular 433, 1980, 1.

851. Morgan, R. P., II, Rasin, V. J., Jr., and Noe, L. A., Sediment effects on eggs and larvae of striped bass and white perch, *Trans. Am. Fish. Soc.*, 112, 220, 1983.

852. Rees, R. A. and Cook, S. F., Effects of sunlight intensity on survival of striped bass X white bass fry, *Proc. S.E. Assoc. Fish Wildl. Agen.*, 36, 83, 1982.

853. Kerby, J. H., Cryogenic preservation of sperm from striped bass, *Trans. Am. Fish. Soc.*, 112, 86, 1983.

854. Gray, L. D., Striped bass for Arkansas?, *Proc. S.E. Assoc. Game Fish Comm.*, 11, 287, 1958.

855. McHugh, J. J. and Heidinger, R. C., Effects of light shock and handling shock on striped bass fry, *Prog. Fish-Cult.*, 40, 82, 1978.

856. Doroshev, S. I., Cornacchia, J. W., and Hogan, K., Initial swim bladder inflation in the larvae of physoclistus fishes and its importance in larval culture, *Rapp. P.-V. Reun. Cong. Int. Explor. Mer.*, 178, 485, 1981.

857. Hawke, J. P., A survey of the diseases of striped bass, *Morone saxatilis* and pompano *Trachinotus carolinus* cultured in earthen ponds, *Proc. World Maricult. Soc.*, 7, 495, 1976.

858. Bayless, J. D., personal communication, 1981.

859. Mitchell, A. J., Parasites and diseases of striped bass, in The Aquaculture of Striped Bass: A Proceedings, McCraren, J. P., Ed., University of Maryland Sea Grant Publ. UM-SG-MAP-84-01, College Park, 1984, 177.

860. Paperna, I. and Zwerner, D. E., Parasites and diseases of striped bass, *Morone saxatilis* (Walbaum), from the lower Chesapeake Bay, *J. Fish Biol.*, 9, 267, 1976.

861. Wellborn, T. L., Jr., The toxicity of nine therapeutic and herbicidal compounds to striped bass, *Prog. Fish-Cult.*, 31, 27, 1969.

862. Wellborn, T. L., Jr., Toxicity of some compounds to striped bass fingerlings, *Prog. Fish-Cult.*, 33, 32, 1971.

863. Schnick, R. A., Meyer, F. P., and Van Meter, H. D., Compounds registered for fisheries uses, *Fisheries*, 4, 18, 1979.

864. Neils, K. E., Commercial production and marketing: striped bass fingerlings, *Comm. Fish Farmer*, 5(2), 18, 1979.

865. Smith, T. I. J., personal communication, 1984.

866. Kerby, J. H., Burrell, V. G., Jr., and Richards, C. E., Occurrence and growth of striped bass X white bass hybrids in the Rappahannock River, Virginia, *Trans. Am. Fish. Soc.*, 100, 787, 1971.

867. Courtesy of Jack D. Bayless, South Carolina Wildlife and Marine Resources Department.

868. Courtesy of R. David Bishop, Tennessee Wildlife Resources Agency.

869. Martin, J. M., The minnow farming industry, in *Proc. Commercial Bait Fish Conf.*, Texas A&M University, College Station, 1968, 5.

870. Giudice, J. J., Gray, D. L., and Martin, J. M., Manual for Bait Fish Culture in the South, University of Arkansas Cooperative Extension Service and U.S. Fish and Wildlife Service, 1981, 1.

871. Giudice, J. J., The culture of bait fishes, in *Proc. Commercial Bait Fish Conf.*, Texas A&M University, College Station, 1968, 13.

872. Flickinger, S. A., Pond Culture of Bait Fishes, Colorado Cooperative Extension Service, Fort Collins, 1971, 1.

873. Martin, J. M., Goldfish Farming, U.S. Fish and Wildlife Service, 1982, 1.

874. Yoshitaka, A., Goldfish Culture, in *Modern Methods of Aquaculture in Japan,* Elsevier, New York, 1983, 79.

875. Smith, E. R., Minnow pond construction and water quality, in *Proc. Commercial Bait Fish Conf.*, Texas A&M University, College Station, 1968, 7.

876. Murai, T. and Andrews, J. W., Effects of salinity on the eggs and fry of the golden shiner and goldfish, *Prog. Fish-Cult.,* 39, 121, 1977.
877. Saylor, M. L., Effect of harvesting methods on production of fingerling fathead minnows, *Prog. Fish-Cult.,* 35, 110, 1973.
878. Nagel, T., Technique for collecting newly hatched fathead minnow fry, *Prog. Fish-Cult.,* 38, 137, 1976.
879. Guest, W. C., Technique for collecting and incubating eggs of the fathead minnow, *Prog. Fish-Cult.,* 39, 188, 1977.
880. Benoit, D. A. and Carlson, R. W., Spawning success of fathead minnows on selected artificial substrates, *Prog. Fish-Cult.,* 39, 67, 1977.
881. Stacey, N. E., Cook, A. F., and Peter, R. E., Spontaneous and gonadotropin-induced ovulation in the goldfish *Carassius auratus* L.: effects of external factors, *J. Fish Biol.,* 15, 349, 1979.
882. Dupree, H. K. and Huner, J. V., Third Report to the Fish Farmers, U.S. Fish and Wildlife Service, 1984, 1.
883. Johnson, S. K., Maintaining Minnows — A Guide for Retailers, Texas Agriculture Extension Service Bulletin B-1365, 1981, 1.
884. Harry, G., 1968, Handling and transporting golden shiner minnows, in *Proc. Commercial Bait Fish Conf.,* Texas A&M University, College Station, 1968, 37.
885. Bishop, H., Parasites and diseases of common bait fishes, in *Proc. Commercial Bait Fish Conf.,* Texas A&M University, College Station, 1968, 31.
886. Goodman, R., personal communication, 1984.

INDEX

McDonald hatching jar, 130, 136
Meat meal, 36, 86
Mechanical grader, 157
Mercury, 33
Metazoan parasite, 146
Methemoglobinemia, 14—15, 27
    treatment of, 28
Micropterus
    dolomieu, see Smallmouth bass
    punctulatus, see Spotted bass
    salmoides, see Largemouth bass
    treculi, see Guadalupe bass
Milk scale disease, 157
Milo pellet, 137
Milt
    of carp, 50
    of yellow perch, 107
Mineral premix, 99
Mineral requirement, 17
    of channel catfish, 38
Minnow, 2, 18
Mirror carp, 44—45, 51
Mitraspora, 157
Monk, see Kettle
Morone
    chrysops, see White bass
    saxitilis, see Striped bass
Mosquito larvae, 71
Mouthbrooder, 65—69
MS-222, 84—85, 93, 145
Mud carp, in polyculture, 49
Mudfish, 68
Mugil cephalus, see Mullet
Mulberry leaf meal, 71
Mullet, in polyculture, 24, 49, 64
Muskellunge, 2, 92—101
    broodfish, 92—93
    cannibalism in, 96—98
    care of sac-fry of, 94—95
    diseases of, 97—98
    dissolved oxygen requirement of, 100
    feeding regime for, 98—100
    fingerling production in, 96—101
    genetics of, 92—95
    hatching of eggs in, 93—95
    intensive culture of, 97—101
    pond culture of, 96—97
    raceway culture of, 101
    reproduction in, 92—95
    spawning of, 92—94
    stocking density of, 97
    tank culture of, 97—101
    temperature tolerance of, 97
Mylopharyngodon piceus, see Black carp
Myxobacteria, 72, 123
Myxobolus
    argentus, 157
    notemigoni, 157

N

Natural food, 16

for common carp, 52—53
    for grass carp, 53
    for tilapia, 69—70
    for walleye, 118
Nest
    for channel catfish, 30—31
    for crappie, 88
    for fathead minnow, 153
    for largemouth bass, 76—77
    for smallmouth bass, 85—86
Net pen culture, of striped bass, 141
Nile perch, 2, 68
Nitrate, 13
Nitrite, 14—15
    removal of, 11
    tolerance of, 14—15
        of bluegill, 89
        of channel catfish, 27—28
        of largemouth bass, 89
        of striped bass, 143—144
Nitrobacter, 11
Nitromonas, 11
Noise, 145
Northern pike, 2, 92—101
    broodfish, 92—93
    cannibalism in, 96—98
    care of sac-fry of, 94—95
    diseases of, 97—98
    feeding regime for, 98—100
    fingerling production in, 96—101
    genetics of, 92—95
    hatching of eggs of, 93—95
    intensive culture of, 97—101
    pond culture of, 96—97
    reproduction in, 92—95
    spawning of, 92—94
    stocking density of, 97
    tank culture of, 97—101
    temperature tolerance of, 97
Notemigonus crysoleucas, see Golden shiner
Notonectidae, 140
Notropis, see Shiner
Nursery pond
    for bigmouth buffalo, 55
    for carp, 50—51
    for tilapia, 63
Nutritional requirements, 16—17
    of baitfish, 154—156
    of channel catfish, 35—38
    of common carp, 53
    of striped bass, 139—140
    of tilapia, 69—72

O

Off-flavor problem, 18
    in channel catfish, 40—41

Z